# Food, Fermentation and Micro-organisms

# Food, Fermentation and Micro-organisms

Charles W. Bamforth
University of California Davis,
USA

**Blackwell**
Science

Editorial offices:
Blackwell Publishing Ltd, 9600 Garsington Road, Oxford OX4 2DQ, UK
    Tel: +44 (0)1865 776868
Blackwell Publishing Professional, 2121 State Avenue, Ames, Iowa 50014-8300, USA
    Tel: +1 515 292 0140
Blackwell Science Asia Pty Ltd, 550 Swanston Street, Carlton, Victoria 3053, Australia
    Tel: +61 (0)3 8359 1011

First published 2005

Library of Congress Cataloging-in-Publication Data
Bamforth, Charles W., 1952–
    Food, fermentation and micro-organisms / Charles W. Bamforth.
        p. cm.
    Includes bibliographical references and index.
    ISBN-13: 978-0-632-05987-4 (hardback: alk. paper)
    ISBN-10: 0-632-05987-7 (hardback: alk. paper)
1. Fermentation. 2. Fermented foods. 3. Yeast.
    [DNLM: 1. Fermentation. 2. Food Microbiology. 3. Alchoholic Beverages - - Microbiology.
    QW 85 B199 2005] I. Title.

    QR151.B355 2005
    664'.024–dc22                                                          2005003336

ISBN-13: 978-0632-05987-4
ISBN-10: 0-632-05987-7

A catalogue record for this title is available from the British Library

Set in 10/13pt TimesNRMT
by Newgen Imaging Systems (P) Ltd, Chennai.

For further information on Blackwell Publishing, visit our website:
www.blackwellpublishing.com

*In honour of Peter Large: scientist, mentor, beer lover, colleague, friend*

*God made yeast, as well as dough, and loves fermentation just as dearly as he loves vegetation.*

Ralph Waldo Emerson (1803–1882)

# Contents

# Preface

I am often asked if I like my job as Professor of Brewing in sunny California, an hour from San Francisco, an hour to the hills, gloriously warm, beautiful people. Does a duck like water? Do round pegs insert into round holes?

But surely, my inquisitors continue, there must be things you miss from your native England? Of course, there are. Beyond family I would have high on the list *The Times*, Wolverhampton Wanderers, truly excellent Indian restaurants and the pub.

If only I could transport one of my old West Sussex locals to downtown Davis! It wouldn't be the same, of course. So I am perforce to reminisce nostalgically.

The beautifully balanced, low carbonation, best bitter ale in a jugged glass. Ploughman's lunches of ham, salami, cheese, pickled onions and freshly baked crusty bread. The delights of the curry, with nan and papadom, yoghurty dips. Glasses of cider or the finest wine (not necessarily imported, but usually). And the rich chocolate pud. Perhaps a post-prandial port, or Armagnac, or Southern Comfort (yes, I confess!).

Just look at that list. Ralph Waldo Emerson hit the nail on the head: what a gift we have in fermentation, the common denominator between all these foodstuffs and many more besides. In this book I endeavour to capture the essence of these very aged and honourable biotechnologies for the serious student of the topic. It would be impossible in a book of this size to do full justice to any of the individual food products – those seeking a fuller treatment for each are referred to the bibliography at the end of each treatment. Rather I seek to demonstrate the clear overlaps and similarities that sweep across all fermented foods, stressing the essential basics in each instance.

# Acknowledgements

I thank my publishers Blackwell, especially Nigel Balmforth and Laura Price, for their patience in awaiting a project matured far beyond its born-on date.

Thanks to Linda Harris, John Krochta, Ralph Kunkee, David Mills and Terry Richardson for reading individual chapters of the book and ensuring that I approach the straight and narrow in areas into which I have strayed from my customary purview. Any errors are entirely my responsibility. One concern is the naming of micro-organisms. Taxonomists seem to be forever updating the Latin monikers for organisms, while the practitioners in the various industries that use the organisms tend to adhere to the use of older names. Thus, for example, many brewers of lager beers in the world still talk of *Saccharomyces carlsbergensis* or *Saccharomyces uvarum* despite the yeast taxonomists having subsequently taken us through *Saccharomyces cerevisiae* lager-type to *Saccharomyces pastorianus*. If in places I am employing an outmoded name, the reader will please forgive me. Those in search of the current 'taxonomical truth' can check it out at http://www.ncbi.nlm.nih.gov/Taxonomy/taxonomyhome.html.

Many thanks to Claudia Graham for furnishing the better drawings in this volume.

And thanks as always to my beloved wife and family: Diane, Peter (and his bride Stephanie), Caroline and Emily.

# Introduction

Campbell-Platt defined fermented foods as 'those foods that have been subjected to the action of micro organisms or enzymes so that desirable biochemical changes cause significant modification in the food'. The processes may make the foods more nutritious or digestible, or may make them safer or tastier, or some or all of these.

Most fermentation processes are extremely old. Of course, nobody had any idea of what was actually happening when they were preparing these products – it was artisan stuff. However, experience, and trial and error, showed which were the best techniques to be handed on to the next generation, so as to achieve the best end results. Even today, some producers of fermented products – even in the most sophisticated of areas such as beer brewing – rely very much on 'art' and received wisdom.

Several of the products described in this book originate from the Middle East (the Fertile Crescent – nowadays known as Iraq) some 10 000–15 000 years ago. As a technique, fermentation was developed as a low energy way in which to preserve foods, featuring alongside drying and salting in days before the advent of refrigeration, freezing and canning. Perhaps the most widespread examples have been the use of lactic acid bacteria to lower the pH and the employment of yeast to effect alcoholic fermentations. Preservation occurs by the conversion of carbohydrates and related components to end products such as acids, alcohols and carbon dioxide. There is both the removal of a prime food source for spoilage organisms and also the development of conditions that are not conducive to spoiler growth, for example, low pH, high alcohol and anaerobiosis. The food retains ample nutritional value, as degradation is incomplete. Indeed changes occurring during the processes may actually increase the nutritional value of the raw materials, for example, the accumulation of vitamins and antioxidants or the conversion of relatively indigestible polymers to more assimilable degradation products.

The crafts were handed on within the home and within feudal estates or monasteries. For the most part batch sizes were relatively small, the production being for local or in-home consumption. However, the Industrial Revolution of the late eighteenth Century led to the concentration of people in towns and cities. The working classes now devoted their labours to work in increasingly heavy industry rather than domestic food production. As a consequence, the fermentation-based industries were focused in fewer larger companies in each sector. Nowadays there continues to be an interest in commercial products produced on the very small scale, with some convinced that such products are superior to those generated by mass production, for example, boutique beers from the brewpub and breads baked in the street corner bakery. More often than not, for beer if not necessarily for bread, this owes more to hype and passion rather than true superiority. Often the converse is true, but it is nonetheless a charming area.

Advances in the understanding of microbiology and of the composition of foods and their raw materials (e.g. cereals, milk), as well as the development of tools such as artificial refrigeration and the steam engine, allowed more consistent processing, while simultaneously vastly expanding the hinterland for each production facility. The advances in microbiology spawned starter cultures, such that the fermentation was able to pursue a predictable course and no longer one at the whim or fancy of indigenous and adventitious microflora.

Thus, do we arrive at the modern day food fermentation processes. Some of them are still quaint – for instance, the operations surrounding cocoa fermentation. But in some cases, notably brewing, the technology in larger companies is as sophisticated and highly controlled as in any industry. Indeed, latter day fermentation processes such as those devoted to the production of pharmaceuticals were very much informed by the techniques established in brewing.

Fermentation in the strictest sense of the word is anaerobic, but most people extend the use of the term to embrace aerobic processes and indeed related non-microbial processes, such as those effected by isolated enzymes.

In this book, we will address a diversity of foodstuffs that are produced according to the broadest definitions of fermentation. I start in Chapter 1 by considering the underpinning science and technology that is common to all of the processes. Then, in Chapter 2, we give particularly detailed attention to the brewing of beer. The reader will forgive the author any perceived prejudice in this. The main reason is that by consideration of this product (from a fermentation industry that is arguably the most sophisticated and advanced of all of the ones considered in this volume), we address a range of issues and challenges that are generally relevant for the other products. For instance, the consideration of starch is relevant to the other cereal-based foods, such as bread, sake and, of course, distilled grain-based beverages. The discussion of *Saccharomyces* and the impact of its metabolism on flavour are pertinent for wine, cider and other alcoholic beverages. (Table 1 gives a summary of the main alcoholic beverages and their relationship to the chief sources of carbohydrate that represent fermentation feedstock.) We can go further: one of the finest examples of vinegar (malt) is fundamentally soured unhopped beer.

The metabolic issues that are started in Chapter 1 and developed in Chapter 2 will inform all other chapters where microbes are considered. Thus, from these two chapters, we should have a well-informed grasp of the generalities that will enable consideration of the remaining foods and beverages addressed in the ensuing chapters.

**Table 1** The relationship between feedstock, primary fermentation products and derived distillation products.

| Raw material | Non-distilled fermentation product | Distilled fermentation derivative |
|---|---|---|
| Apple | Cider | Apple brandy, Calvados |
| Barley | Beer | Whisk(e)y |
| Cacti/succulents | Pulque | Tequila |
| Grape | Wine | Armagnac, Brandy, Cognac |
| Palmyra | Toddy | Arak |
| Pear | Perry | Pear brandy |
| Honey | Mead | |
| Rice | Sake | Shochu |
| Sorghum | Kaffir beer | |
| Sugar cane/molasses | | Rum |
| Wheat | Wheat beer | |

Whisky is not strictly produced by distillation of beer, but rather from the very closely related fermented unhopped wash from the mashing of malted barley.

# Bibliography

Angold, R., Beech, G. & Taggart, J. (1989) *Food Biotechnology: Cambridge Studies in Biotechnology 7*. Cambridge: Cambridge University Press.

Caballero, B., Trugo, L.C. & Finglas, P.M., eds (2003) *Encyclopaedia of Food Sciences and Nutrition*. Oxford: Academic Press.

Campbell-Platt, G. (1987) *Fermented Foods of the World: A Dictionary and Guide*. London: Butterworths.

King, R.D. & Chapman, P.S.J., eds (1988) *Food Biotechnology*. London: Elsevier.

Lea, A.G.H. & Piggott, J.R., eds (2003) *Fermented Beverage Production*, 2nd edn. New York: Kluwer/Plenum.

Peppler, H.J. & Perlman, D., eds (1979) *Microbial Technology*. New York: Academic Press.

Reed, G., ed (1982) *Prescott and Dunn's Industrial Microbiology*, 4th edn. Westport, CT: AVI.

Rehm, H.-J. & Reed, G., eds (1995) *Biotechnology*, 2nd edn, vol. 9, Enzymes, Biomass, Food and Feed. Weinheim: VCH.

Rose, A.H., ed. (1977) *Alcoholic Beverages*. London: Academic Press.

Rose, A.H., ed. (1982a) *Economic Microbiology*. London: Academic Press.

Rose, A.H., ed. (1982b) *Fermented Foods*. London: Academic Press.

Varnam, A.H. & Sutherland, J.P. (1994) *Beverages: Technology, Chemistry and Microbiology*. London: Chapman & Hall.

Wood, B.J.B., ed. (1998) *Microbiology of Fermented Foods*, 2nd edn, 2 vols. London: Blackie.

# Chapter 1
# The Science Underpinning Food Fermentations

Use the word 'biotechnology' nowadays and the vast majority of people will register an image of genetic alteration of organisms in the pursuit of new applications and products, many of them pharmaceutically relevant. Even the *Merriam-Webster's Dictionary* tells me that biotechnology is 'biological science when applied especially in genetic engineering and recombinant DNA technology'. Fortunately, the *Oxford English Dictionary* gives a rather more accurate definition as 'the branch of technology concerned with modern forms of industrial production utilising living organisms, especially microorganisms, and their biological processes'.

Accepting the truth of the second of these, we can realise that biotechnology is far from being a modern concept. It harks back historically vastly longer than the traditional milepost for biotechnology, namely Watson and Crick's announcement in the Eagle pub in Cambridge (and later, more formally, in *Nature*) that they had found 'the secret of life'.

Eight thousand years ago, our ancient forebears may have been, in their own way, no less convinced that they had hit upon the essence of existence when they made the first beers and breads. The first micro-organism was not seen until draper Anton van Leeuwenhoek peered through his microscope in 1676, and neither were such agents firmly causally implicated in food production and spoilage until the pioneering work of Needham, Spallanzani and Pasteur and Bassi de Lodi in the eighteenth and nineteenth centuries.

Without knowing the whys and wherefores, the dwellers in the Fertile Crescent (nowadays Iraq) were the first to have made use of living organisms in fermentation processes. They truly were the first biotechnologists. And so, beer, bread, cheese, wine and most of the other foodstuffs being considered in this book come from the oldest of processes. In some cases these have not changed very much in the ensuing aeons.

Unlike the output from modern biotechnologies, for the most part, we are considering high volume, low-value commodities. However, for products such as beer, there is now a tremendous scientific understanding of the science that underpins the product, science that is none the less tempered with the pressures of tradition, art and emotion. For all of these food fermentation products, the customer *expects*. As has been realised by those who would apply molecular biological transformations to the organisms involved in the manufacture of foodstuffs, there is vastly more resistance to this than for applications in, say, the pharmaceutical area. You do not mess with a person's meal.

Historically, of course, the micro-organisms employed in these fermentation processes were adventitious. Even then, however, it was realised that the addition of a part of the previous process stream to the new batch could serve to 'kick off' the process. In some businesses, this was called 'back slopping'. We now know that what the ancients were doing was seeding the process with a hefty dose of the preferred organism(s). Only relatively recently have the relevant microbes been added in a purified and enriched form to knowingly seed fermentation processes.

The two key components of a fermentation system are the organism and its feedstock. For some products, such as wine and beer, there is a radical modification of the properties of the feedstock, rendering them more palatable (especially in the case of beer: the grain extracts pre-fermentation are most unpleasant in flavour; by contrast, grape juice is much more acceptable). For other products, the organism is less central, albeit still important. One thinks, for instance, of bread, where not all styles involve yeast in their production.

For products such as cheese, the end product is quite distinct from the raw materials as a result of a series of unit operations. For products such as beer, wine and vinegar, our product is actually the spent growth medium – the excreta of living organisms if one had to put it crudely. Only occasionally is the product the actual micro-organism itself – for example, the surplus yeast generated in a brewery fermentation or that generated in a 'single-cell protein' operation such as mycoprotein.

Organisms employed in food fermentations are many and diverse. The key players are lactic acid bacteria, in dairy products for instance, and yeast, in the production of alcoholic beverages and bread. Lactic acid bacteria, to illustrate, may also have a positive role to play in the production of certain types of wines and beers, but equally they represent major spoilage organisms for such products. It truly is a case of the organism being in the right niche for the product in question.

In this chapter, I focus on the generalities of science and technology that underpin fermentations and the organisms involved. We look at commonalities in terms of quality, for example, the Maillard reaction that is of widespread significance as a source of colour and aroma in many of the foods that we consider. The reader will discover (and this betrays the primary expertise of the author) that many of the examples given are from beer making. It must be said, however, that the scientific understanding of the brewing of beer is somewhat more advanced than that for most if not all of the other foodstuffs described in this book. Many of the observations made in a brewing context translate very much to what must occur in the less well-studied foods and beverages.

## Micro-organisms

Microbes can be essentially divided into two categories: the prokaryotes and the eukaryotes. The former, which embrace the bacteria, are substantially the simpler, in that they essentially comprise a protective cell wall, surrounding a plasma membrane, within which is a nuclear region immersed in cytoplasm (Fig. 1.1). This is a somewhat simplistic description, but suitable for our needs. The nuclear material (deoxyribonucleic acid, DNA), of course, figures as the genetic blueprint of the cell. The cytoplasm contains the enzymes that catalyse the reactions necessary for growth, survival and reproduction of the organisms (the sum total of reactions, of course, being referred to as *metabolism*). The membrane regulates the entry and exit of materials into and from the cell.

The eukaryotic cell (of which baker's or brewer's yeast, *Saccharomyces cerevisiae*, a unicellular fungus, is the model organism) is substantially more complex (Fig. 1.2). It is divided into organelles, the intracellular equivalent

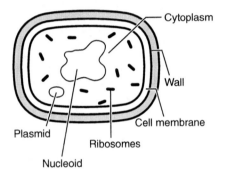

**Fig. 1.1**  A simple representation of a prokaryotic cell. The major differences between Gram-positive and Gram-negative cells concern their outer layers, with the latter having an additional membrane outwith the wall in addition to a different composition in the wall itself.

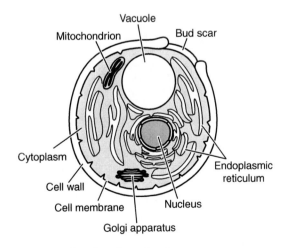

**Fig. 1.2**  A simple representation of a eukaryotic cell.

of our bodily organs. Each has its own function. Thus, the DNA is located in the nucleus which, like all the organelles, is bounded by a membrane. All the membranes in the eukaryotes (and the prokaryotes) comprise lipid and protein. Other major organelles in eukaryotes are the mitochondria, wherein energy is generated, and the endoplasmic reticulum. The latter is an interconnected network of tubules, vesicles and sacs with various functions including protein and sterol synthesis, sequestration of calcium, production of the storage polysaccharide glycogen and insertion of proteins into membranes. Both prokaryotes and eukaryotes have polymeric storage materials located in their cytoplasm.

Table 1.1 lists some of the organisms that are mentioned in this book. Some of the relevant fungi are unicellular, for example, Saccharomyces. However, the major class of fungi, namely the filamentous fungi with their hyphae (moulds), are of significance for a number of the foodstuffs, notably those Asian products involving solid-state fermentations, for example, sake and miso, as well as the only successful and sustained single-cell protein operation (see Chapter 17).

**Table 1.1**   Some micro-organisms involved in food fermentation processes.

| Bacteria | | Fungi | |
|---|---|---|---|
| Gram negative[a] | Gram positive[a] | Filamentous | Yeasts and non-filamentous fungi |
| Acetobacter | Arthrobacter | Aspergillus | Brettanomyces |
| Acinetobacter | Bacillus | Aureobasidium | Candida |
| Alcaligenes | Bifidobacterium | Fusarium | Cryptococcus |
| Escherichia | Cellulomonas | Mucor | Debaromyces |
| Flavobacterium | Corynebacter | Neurospora | Endomycopsis |
| | Lactobacillus | Penicillium | Geotrichum |
| Gluconobacter | Lactococcus | Rhizomucor | Hanseniaspora (Kloeckera) |
| Klebsiella | Leuconostoc | Rhizopus | Hansenula |
| Methylococcus | Micrococcus | Trichoderma | Kluyveromyces |
| Methylomonas | Mycoderma | | Monascus |
| Propionibacter | Staphylococcus | | Pichia |
| Pseudomonas | Streptococcus | | Rhodotorula |
| Thermoanaerobium | Streptomyces | | Saccharomyces |
| Xanthomonas | | | Saccharomycopsis |
| Zymomonas | | | Schizosaccharomyces |
| | | | Torulopsis |
| | | | Trichosporon |
| | | | Yarrowia |
| | | | Zygosaccharomyces |

[a] Danish microbiologist Hans Christian Gram (1853–1928) developed a staining technique used to classify bacteria. A basic dye (crystal violet or gentian violet) is taken up by both Gram-positive and Gram-negative bacteria. However, the dye can be washed out of Gram-negative organisms by alcohol, such organisms being counterstained by safranin or fuchsin. The latter stain is taken up by both Gram-positive and Gram-negative organisms, but does not change the colour of Gram-positive organisms, which retain their violet hue.

# Microbial metabolism

In order to grow, any living organism needs a supply of nutrients that will feature as, or go on to form, the building blocks from which that organism is made. These nutrients must also provide the energy that will be needed by the organism to perform the functions of accumulating and assimilating those nutrients, to facilitate moving around, etc.

The microbial kingdom comprises a huge diversity of organisms that are quite different in their nutritional demands. Some organisms (*phototrophs*) can grow using light as a source of energy and carbon dioxide as a source of carbon, the latter being the key element in organic systems. Others can get their energy solely from the oxidation of inorganic materials (*lithotrophs*).

All of the organisms considered in this book are *chemotrophs*, insofar as their energy is obtained by the oxidation of chemical species. Furthermore, unlike the *autotrophs*, which can obtain all (or nearly all) their carbon from carbon dioxide, the organisms that are at the heart of fermentation processes for making foodstuffs are *organotrophs* (or *heterotrophs*) in that they oxidise organic molecules, of which the most common class is the sugars.

## *Nutritional needs*

The four elements required by organisms in the largest quantity (gram amounts) are carbon, hydrogen, oxygen and nitrogen. This is because these are the elemental constituents of the key cellular components of carbohydrates (Fig. 1.3), lipids (Fig. 1.4), proteins (Fig. 1.5) and nucleic acids (Fig. 1.6). Phosphorus and sulphur are also important in this regard. Calcium, magnesium, potassium, sodium and iron are demanded at the milligram level, while microgram amounts of copper, cobalt, zinc, manganese, molybdenum, selenium and nickel are needed. Finally, organisms need a preformed supply of any material that is essential to their well-being, but that they cannot themselves synthesise, namely vitamins (Table 1.2). Micro-organisms differ greatly in their ability to make these complex molecules. In all instances, vitamins form a part of coenzymes and prosthetic groups that are involved in the functioning of the enzymes catalysing the metabolism of the organism.

As the skeleton of all the major cellular molecules (other than water) comprises carbon atoms, there is a major demand for carbon.

Hydrogen and oxygen originate from substrates such as sugars, but of course also come from water.

The oxygen molecule, $O_2$, is essential for organisms growing by aerobic respiration. Although fermentation is a term that has been most widely applied to an anaerobic process in which organisms do not use molecular oxygen in respiration, even those organisms that perform metabolism in this way generally do require a source of this element. To illustrate, a little oxygen is introduced into a brewer's fermentation so that the yeast can use it in reactions that are involved in the synthesis of the unsaturated fatty acids and sterols that

(a)

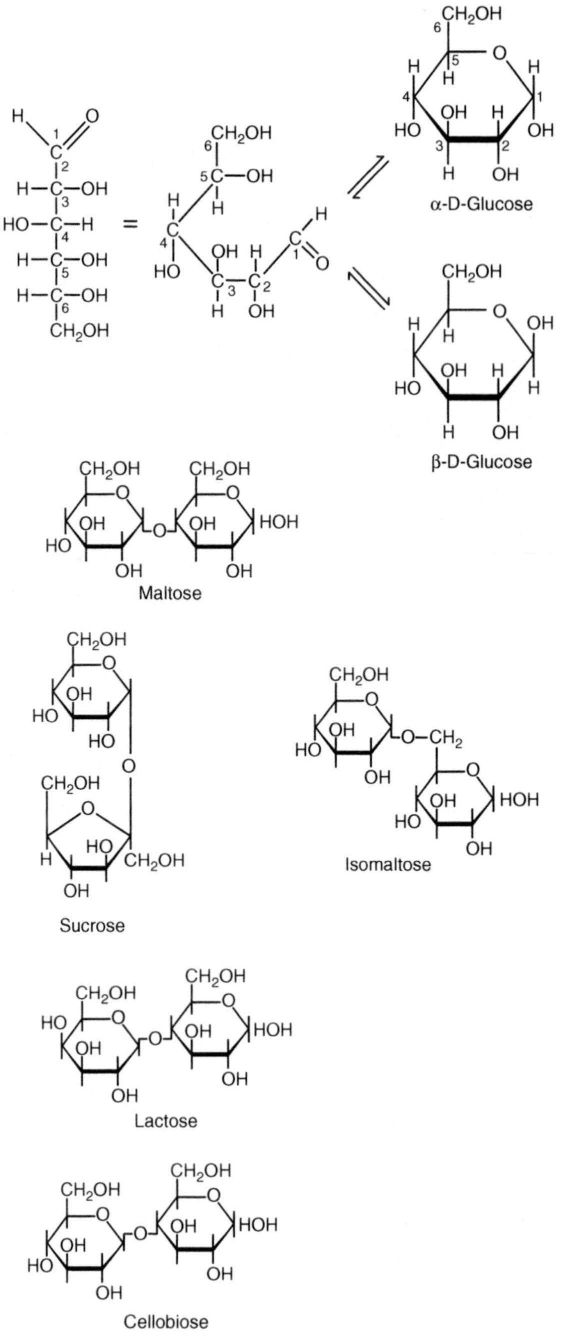

α-D-Glucose

β-D-Glucose

Maltose

Sucrose

Isomaltose

Lactose

Cellobiose

**Fig. 1.3**   (Continued).

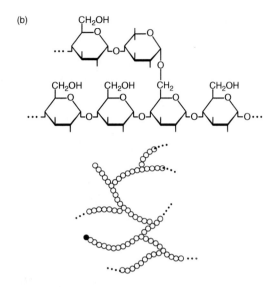

(b)

**Fig. 1.3**  *Carbohydrates.* (a) Hexoses (sugars with six carbons), such as glucose, exist in linear and cyclic forms in equilibria (top). The numbering of the carbon atoms is indicated. In the cyclic form, if the OH at $C_1$ is lowermost, the configuration is $\alpha$. If the OH is uppermost, then the configuration is $\beta$. At $C_1$ in the linear form is an aldehyde grouping, which is a reducing group. Adjacent monomeric sugars (monosaccharides, in this case glucose) can link (condense) by the elimination of water to form disaccharides. Thus, maltose comprises two glucose moieties linked between $C_1$ and $C_4$, with the OH contributed by the $C_1$ of the first glucosyl residue being in the $\alpha$ configuration. Thus, the bond is $\alpha 1 \rightarrow 4$. For isomaltose, the link is $\alpha 1 \rightarrow 6$. For cellobiose, the link is $\beta 1 \rightarrow 4$. Sucrose is a disaccharide in which glucose is linked $\beta 1 \rightarrow 4$ to a different hexose sugar, fructose. Similarly, lactose is a disaccharide in which galactose (note the different conformation at its $C_4$) is linked $\beta 1 \rightarrow 4$ to glucose. (b) Successive condensation of sugar units yields oligosaccharides. This is a depiction of part of the amylopectin fraction of starch, which includes chains of $\alpha 1 \rightarrow 4$ glucosyls linked by $\alpha 1 \rightarrow 6$ bonds. The second illustration shows that there is only one glucosyl (marked by •) that retains a free $C_1$ reducing group, all the others (○) being bound up in glycosidic linkages.

are essential for it to have healthy membranes. Aerobic metabolism, too, is necessary for the production of some of the foodstuffs mentioned in this book, for example, in the production of vinegar.

All growth media for micro-organisms must incorporate a source of nitrogen, typically at $1–2\,\mathrm{g\,L^{-1}}$. Most cells are about 15% protein by weight, and nitrogen is a fundamental component of protein (and nucleic acids).

As well as being physically present in the growth medium, it is equally essential that the nutrient should be capable of entering into the cell. This transport is frequently the rate-limiting step. Few nutrients enter the cell by passive diffusion and those that do tend to be lipid-soluble. Passive diffusion is not an efficient strategy for a cell to employ as it is very concentration-dependent. The rate and extent of transfer depend on the relative concentrations of the substance inside and outside the cell. For this reason, facilitated transportation is a major mechanism for transporting materials (especially water-soluble ones) into the cell, with proteins known as permeases selectively and specifically catalysing the movement. These permeases are only synthesised as and

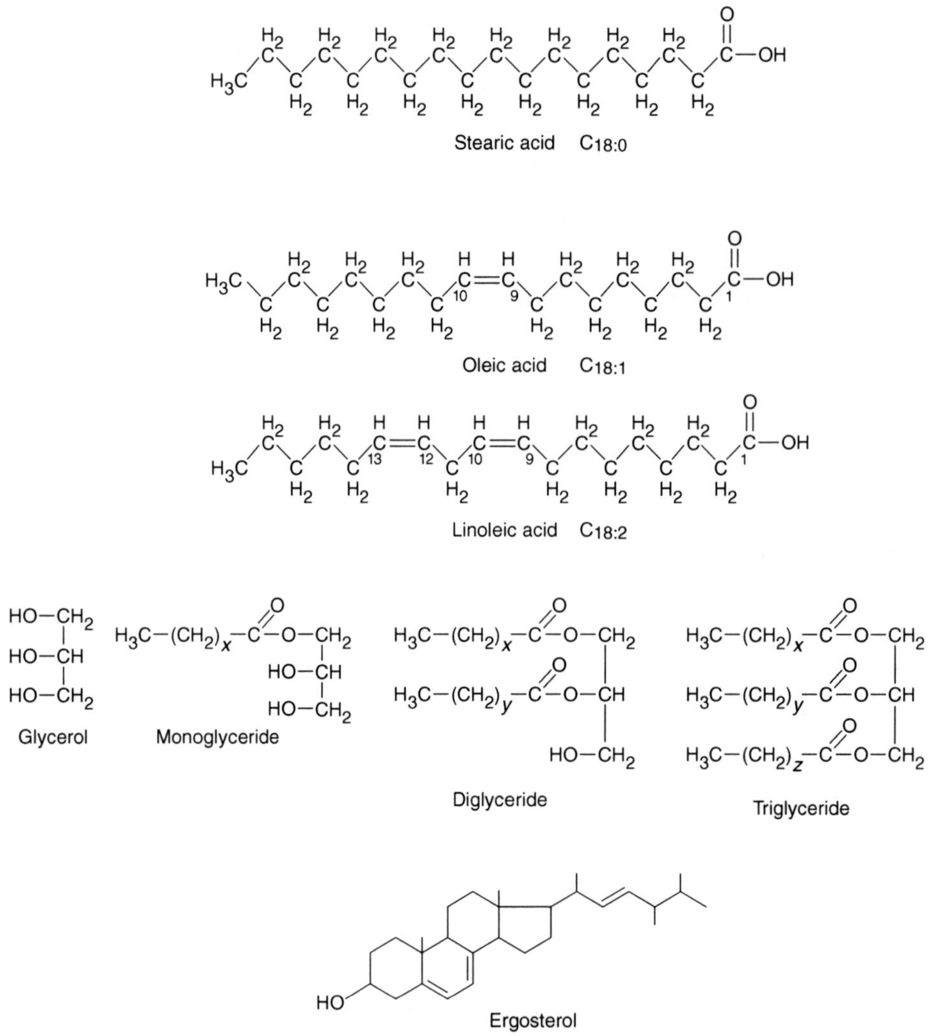

Stearic acid    C$_{18:0}$

Oleic acid    C$_{18:1}$

Linoleic acid    C$_{18:2}$

Glycerol

Monoglyceride

Diglyceride

Triglyceride

Ergosterol

**Fig. 1.4** *Lipids.* Fatty acids comprise hydrophobic hydrocarbon chains varying in length, with a single polar carboxyl group at C$_1$. Three different fatty acids with 18 carbons (hence C$_{18}$) are shown. They are the 'saturated' fatty acid stearic acid (so-called because all of its carbon atoms are linked either to another carbon or to hydrogen with no double bonds) and the unsaturated fatty acids, oleic acid (one double bond, hence C$_{18:1}$) and linoleic acid (two double bonds, C$_{18:2}$). Fatty acids may be in the free form or attached through ester linkages to glycerol, as glycerides.

when the cell requires them. In some instances, energy is expended in driving a substance into the cell if a thermodynamic hurdle has to be overcome, for example, a higher concentration of the molecule inside than outside. This is known as 'active transport'.

An additional challenge is encountered with high molecular weight nutrients. Whereas some organisms, for example, the protozoa, can assimilate these materials by engulfing them (*phagocytosis*), micro-organisms secrete extracellular enzymes to hydrolyse the macromolecule outside the organism, with

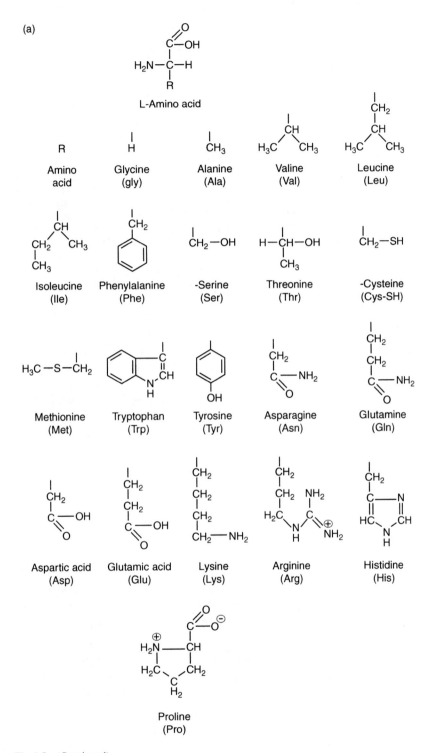

Fig. 1.5 (Continued).

(b)

$$H_3\overset{\oplus}{N}-CH-\overset{O}{\underset{R_1}{\overset{\|}{C}}}-\overset{\ominus}{O} \quad + \quad H_3\overset{\oplus}{N}-CH-\overset{O}{\underset{R_2}{\overset{\|}{C}}}-\overset{\ominus}{O} \quad \underset{+ H_2O}{\overset{- H_2O}{\rightleftharpoons}} \quad H_3\overset{\oplus}{N}-CH-\overset{O}{\underset{R_1}{\overset{\|}{C}}}-NH-CH-\overset{O}{\underset{R_2}{\overset{\|}{C}}}-\overset{\ominus}{O}$$

**Fig. 1.5** *Proteins.* (a) The monomeric components of proteins are the amino acids, of which there are 19 major ones and the imino acid proline. The amino acids have a common basic structure and differ in their R group. The amino groups in the molecules can exist in free ($-NH_2$) and protonated ($-NH_3^+$) forms depending on the pH. Similarly, the carboxyl groups can be in the protonated ($-COOH$) and non-protonated ($-COO^-$) states. (b) Adjacent amino acids can link through the 'peptide' bond. Proteins contain many amino acids thus linked. Such long, high molecular weight molecules adopt complex three-dimensional forms through interactions between the amino acid R groups, such structures being important for the properties that different proteins display.

the resultant lower molecular weight products then being assimilated. These extracellular enzymes are nowadays produced commercially in fermentation processes that involve subsequent recovery of the spent growth medium containing the enzyme and various degrees of ensuing purification. A list of such enzymes and their current applications is given in Table 1.3.

## Environmental impacts

A range of physical, chemical and physicochemical parameters impact the growth of micro-organisms, of which we may consider temperature, pH, water activity, oxygen, radiation, pressure and 'static' agents.

### Temperature

The rate of a chemical reaction was shown by Svante Arrhenius (1859–1927) to increase two- to three-fold for every 10°C rise in temperature. However, cellular macromolecules, especially the enzymes, are prone to denaturation by heat, and this accordingly limits the temperatures that can be tolerated. Although there are organisms that can thrive at relatively high temperatures (*thermophiles*), most of the organisms discussed in this book do not fall into that class. Neither do they tend to be *psychrophiles*, which are organisms capable of growth at very low temperatures. They have a minimum temperature at which growth can occur, below which the lipids in the membranes are insufficiently fluid. It should be noted that many organisms can *survive* (if not grow) at lower temperatures and advantage is taken of this in the storage of pure cultures of defined organisms (discussed later). Organisms which prefer the less-extreme temperature brackets, say 10–40°C, are referred to as *mesophiles*.

(a)

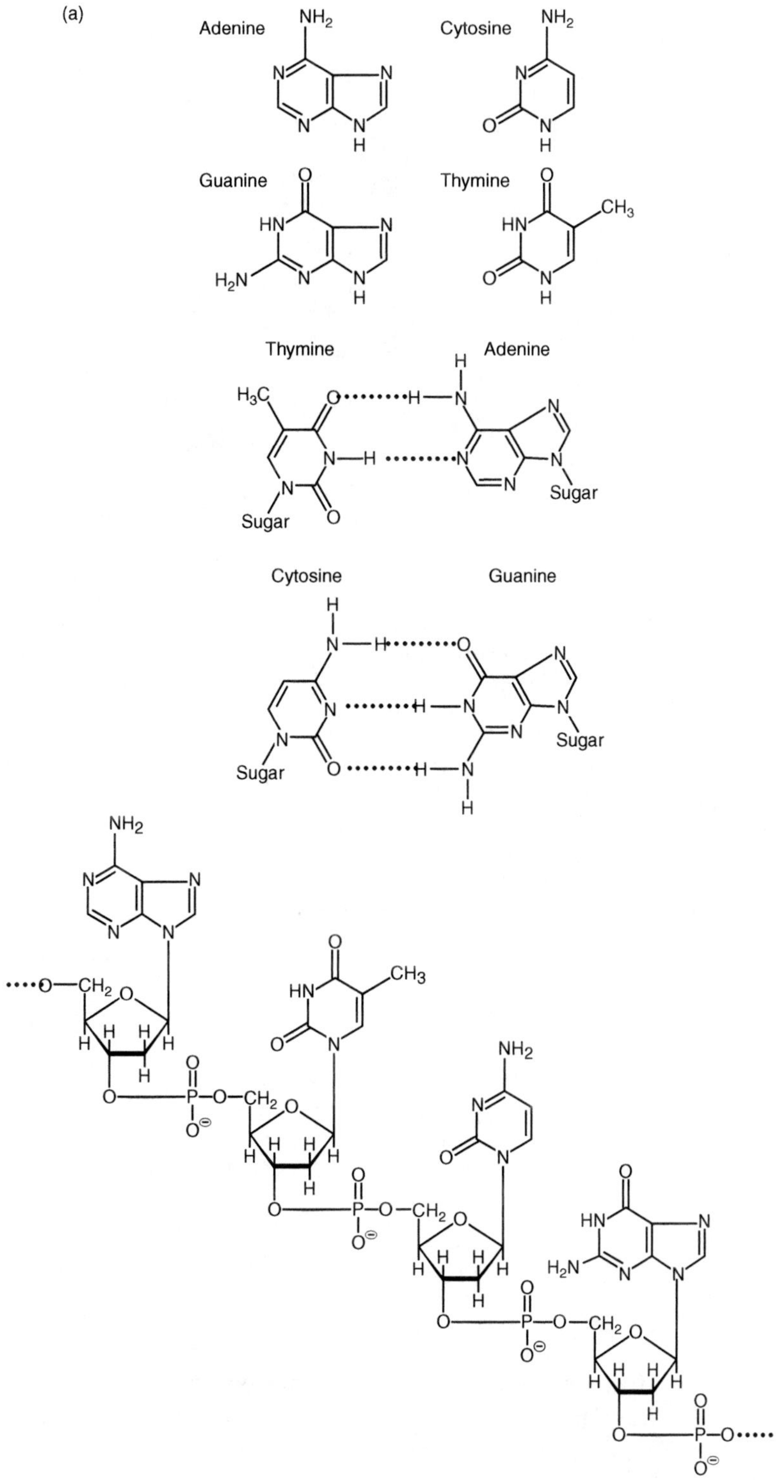

**Fig. 1.6**   (Continued).

(b)

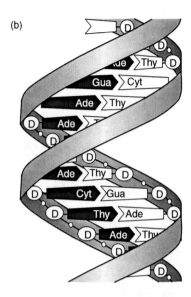

**Fig. 1.6**  *Nucleic acids*. (a) Nucleic acids comprise three building blocks: bases, pentose (sugars with five carbon atoms) and phosphate. There are four bases in DNA: the purines adenine (A) and guanine (G) and the pyrimidines thymine (T) and cytosine (C). A and T or G and C can interact through hydrogen bonds (dotted lines) and this binding affords the linking between adjacent chains in DNA. The bases are linked to the sugar–phosphate backbone. (b) In the famous double-helix form of DNA, adjacent strands of deoxyribose (D)–phosphate (o) are linked through the bases. The sequence of bases represents the genetic code that determines the properties of any living organism. In ribonucleic acid (RNA), there is only one strand: thymine is replaced by another pyrimidine (uracil) and the sugar is ribose, whose $C_2$ has an –OH group rather than two H atoms.

**Table 1.2**  Role of vitamins in micro-organisms.

| Vitamin | Coenzyme it forms part of |
| --- | --- |
| Thiamine (vitamin $B_1$) | Thiamine pyrophosphate |
| Riboflavin ($B_2$) | Flavin adenine dinucleotide, flavin mononucleotide |
| Niacin | Nicotinamide adenine dinucleotide |
| Pyridoxine ($B_6$) | Pyridoxal phosphate |
| Pantothenate | Coenzyme A |
| Biotin | Prosthetic group in carboxylases |
| Folate | Tetrahydrofolate |
| Cobalamin ($B_{12}$) | Cobamides |

## *pH*

Most organisms have a relatively narrow range of pH within which they grow best. This tends to be lower for fungi than it is for bacteria. The optimum pH of the medium reflects the best compromise position in respect of

(1) the impact on the surface charge of the cells (and the influence that this has on behaviours such as flocculation and adhesion);

**Table 1.3**  Exogenous enzymes.

| Enzyme | Major sources | Application in foods |
|---|---|---|
| $\alpha$-Amylase | Aspergillus, Bacillus | Syrup production, baking, brewing |
| $\beta$-Amylase | Bacillus, Streptomyces, Rhizopus | Production of high maltose syrups, brewing |
| Glucoamylase | Aspergillus, Rhizopus | Production of glucose syrups, baking, brewing, wine making |
| Glucose isomerase | Arthrobacter, Streptomyces | Production of high fructose syrups |
| Pullulanase | Klebsiella, Bacillus | Starch (amylopectin) degradation |
| Invertase | Kluyveromyces, Saccharomyces | Production of invert sugar, production of soft-centred chocolates |
| Glucose oxidase (coupled with catalase) | Aspergillus, Penicillium | Removal of oxygen in various foodstuffs |
| Pectinase | Aspergillus, Penicillium | Fruit juice and wine production, coffee bean fermentation |
| $\beta$-Glucanases | Bacillus, Penicillium, Trichoderma | Brewing, fruit juices, olive processing |
| Pentosanases | Cryptococcus, Trichosporon | Baking, brewing |
| Proteinases | Aspergillus, Bacillus, Rhizomucor, Lactococcus, recombinant Kluyveromyces, Papaya | Baking, brewing, meat tenderisation, cheese |
| Catalase | Micrococcus, Corynebacterium, Aspergillus | Cheese (see also glucose oxidase above) |
| Lipases | Aspergillus, Bacillus, Rhizopus, Rhodotorula | Dairy and meat products |
| Urease | Lactobacillus | Wine |
| Tannase | Aspergillus | Brewing |
| $\beta$-Galactosidase | Aspergillus, Bacillus, Escherichia, Kluyveromyces | Removal of lactose |
| Acetolactate decarboxylase | Thermoanaerobium | Accelerated maturation of beer |

(2) on the ability of the cells to maintain a desirable intracellular pH and, in concert with this, the charge status of macromolecules (notably the enzymes) and the impact that this has on their ability to perform.

*Water activity*

The majority of microbes comprise between 70% and 80% water. Maintaining this level is a challenge when an organism is exposed variously to environments that contain too little water (dehydrating or *hypertonic* locales) or excess water (*hypotonic*).

The water that is available to an organism is quantifiable by the concept of water activity ($A_w$). Water activity is defined as the ratio of the vapour pressure of water in the solution surrounding the micro-organism to the vapour pressure of pure water. Thus, pure water itself has an $A_w$ of 1 while an absolutely dry, water-free entity would have an $A_w$ of 0. Micro-organisms differ greatly in the extent to which they will tolerate changes in $A_w$. Most bacteria will not grow below $A_w$ of 0.9, so drying is a valuable means for protecting against spoilage by these organisms. By contrast, many of the fungi that can spoil grain ($A_w = 0.7$) can grow at relatively low moisture levels and are said to be *xerotolerant*. Truly *osmotolerant* organisms will grow at an $A_w$ of 0.6.

*Oxygen*

Microbes differ substantially in their requirements for oxygen. *Obligate aerobes* must have oxygen as the terminal electron acceptor for aerobic growth (Fig. 1.7). *Facultative anaerobes* can use oxygen as terminal electron acceptor, but they can function in its absence. *Microaerophiles* need relatively small proportions of oxygen in order to perform certain cellular activities, but the oxygen exposure should not exceed 2–10% v/v (cf. the atmospheric level of 21% v/v). *Aerotolerant anaerobes* do not use molecular oxygen in their metabolism but are tolerant of it. *Obligate anaerobes* are killed by oxygen.

Clearly these differences have an impact on the susceptibility of foodstuffs to spoilage. Most foods when sealed are (or rapidly become) relatively anaerobic, thus obviating the risk from the first three categories of organism.

Irrespective of which class an organism falls into, oxygen is still a potentially damaging molecule when it becomes partially reduced and converted into

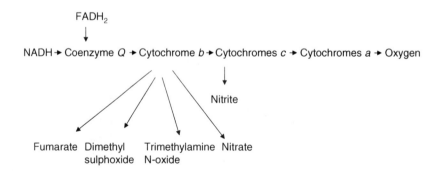

**Fig. 1.7** Electron transport chains. Reducing power captured as NADH or FADH$_2$ is transferred successively through a range of carriers until ultimately reducing a terminal electron acceptor. In aerobic organisms, this acceptor is oxygen, but other acceptors found in many microbial systems are illustrated. This can impact parameters such as food flavour – for example, reduction of trimethylamine N-oxide affords trimethylamine (fishy flavour) while reduction of dimethyl sulphoxide (DMSO) yields dimethyl sulphide (DMS), which is important in the flavour of many foodstuffs.

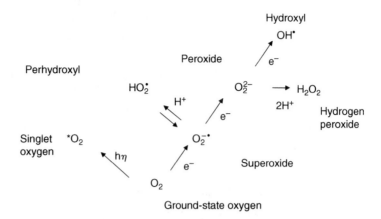

**Fig. 1.8**  Activation of oxygen. Ground-state oxygen is relatively unreactive. By acquiring electrons, it become successively more reactive – superoxide, peroxide, hydroxyl. Superoxide exists in charged and protonated forms, the latter (perhydroxyl) being the more reactive. Exposure to light converts oxygen to another reactive form, singlet oxygen.

radical forms (Fig. 1.8). Organisms that can tolerate oxygen have developed a range of enzymes that scavenge radicals, amongst them superoxide dismutase, catalase and glutathione peroxidase.

## Radiation

One of the radical forms of oxygen, singlet oxygen, is produced by exposure to visible light. An even more damaging segment of the radiation spectrum is the ultraviolet light, exposure to which can lead to damage of DNA. Ionising radiation, such as gamma rays, causes the production of an especially reactive oxygen derived radical, hydroxyl (OH•), and one of the numerous impacts of this is the breakage of DNA. Thus, radiation is a very powerful technique for removing unwanted microbes, for example, in food treatment operations.

## Hydrostatic pressure

In nature, many microbes do not encounter forces exceeding atmospheric pressure (1 atm = 101.3 kPa = 1.013 bar). Increasing the pressure tends to at least inhibit if not destroy an organism. Pressure is of increasing relevance in food fermentation systems because modern fermenters hold such large volumes that pressure may exceed 1.5 atm in some instances. Although they do not necessarily kill organisms, high pressures do impact how organisms behave, including their tendency to aggregate and certain elements of their metabolism. The latter is at least in part due to the accumulation of carbon dioxide that occurs when pressure is increased.

## Controlling or inhibiting growth of micro-organisms

It is important to regulate those organisms that are present during the making of fermentation products and also those that are able to grow and survive in the finished product. On the one hand, we have nowadays the deliberate seeding of the desired organism(s), which therefore gain a selective advantage in outgrowing other organisms. Conversely, there are physical or chemical '-cidal' treatments or sterilisation procedures that are employed to achieve the depletion or total kill of organisms.

Relevant factors are

(1) how many organisms are present;
(2) the types of organism that are present;
(3) the concentration of antimicrobial agents that are present or the intensity of the physical treatment;
(4) the prevailing conditions of temperature, pH and viscosity;
(5) the period of exposure; and
(6) the concentration of organic matter.

Fermentation by itself comprises a procedure that originally emerged as a means for preserving the nutritive value of foodstuffs. Through fermentation there was either the lowering of levels of substances that contaminating organisms would need to support their growth or the development of materials or conditions that would prevent organisms from developing, for example, a lowering of pH. In the case of a product like beer, there is the deliberate introduction of antiseptic agents, in this case, the bitter acids from hops.

### Heating

Moist heat is used for sterilising a greater diversity of materials than dry heat. Moist heat employs steam under pressure and is very effective for the sterilisation of production vessels and pipe work. Dry heat is less efficient and requires a higher temperature (e.g. 160°C as opposed to 120°C); it is used in systems like glassware and for moisture-sensitive materials.

The microbial content of finished food products is frequently lowered by heat treatment. Ultra-high temperature (UHT) treatments are used where especially high kills are necessary. Pasteurisation is a milder process, one in which the temperature and the time of exposure are regulated to achieve a sufficient kill of spoilage organisms without deleteriously impacting the other properties of the foodstuff. In batch pasteurisation, filled containers (e.g. bottles of beer) are held at, say, 62°C for 10 min in chambers through which the product slowly passes on a conveyor (tunnel pasteurisation). In flash pasteurization, the liquid is heated as it flows through heat exchangers en route to the packaging operation. Residence times are much shorter so temperatures are higher (e.g. 72°C for 15 s). In the specific example of beer, this might be the way in which beer destined for kegs is processed. One pasteurisation unit (PU) is defined as exposure to 60°C for 1 min. As the temperature is increased, the shorter exposure time equates to 1 PU. The more organisms, the more extensive is the heat treatment, so the onus is on the operator to minimise the populations by good hygienic practice.

### Cooling

The ability of organisms to grow is curtailed as the temperature is lowered (refrigeration, freezing).

### Drying

As organisms usually require significant amounts of water (discussed earlier), drying affords preservation. Thus, for example, starting materials for fermentation (such as grains and fruits) may be subjected to some degree of drying if they are to be stored successfully prior to use. The other way in which water activity can be lowered is by adding solutes such as salt or sugar. In this book, we encounter several instances where there is deliberate salting during processing to achieve food preservation, for example, in fermented fish production.

### Irradiation

The use of irradiation to eliminate spoilage organisms is charged with emotion. Critics hit on the tendency of the technique to reduce the food value, for example, by damaging vitamins. However, the procedure really should be considered on a case-by-case basis, and only if there is some definite negative impact on the quality of a product should it necessarily be avoided. Thus, to take beer as our example again, there is evidence for the increased production of hydrogen sulphide when beer is irradiated.

### Filtration

Undesirable organisms can be removed by physically filtering them from the product. Depth filters operate by trapping and adsorbing the cells in a fibrous or granular matrix. Membrane filters possess defined pore sizes through which organisms of greater dimensions cannot pass. Typically these pore sizes may be $0.45\,\mu m$ or, for especially rigorous 'clean-up', $0.2\,\mu m$. Practical systems may employ successive filters – for example, a depth filter followed by membranes of different sizes. The approach may be most valuable for heat-sensitive products.

### Chemical agents

Modern food production facilities are designed so that they are readily cleanable between production runs by chemical treatment regimes, often called 'cleaning in place' or CIP. This demands fabrication with resilient material, for example, stainless steel, as well as design that ensures that the agent reaches all nooks and crannies. CIP protocols generally involve an initial water rinse to remove loose soil, followed by a 'detergent' wash. This is not so much a detergent proper as sodium hydroxide or nitric acid and it is targeted at tougher adhering materials. Next is another water rinse to eliminate the detergent, followed by a sterilant. Various chemical sterilants are available, the most commonly used being chlorine, chlorine dioxide and peracetic acid.

Some foodstuffs are formulated so that they contain preservatives (Table 1.4). In other foodstuffs there are natural antimicrobial compounds present, for example, polyphenols and the hop iso-$\alpha$-acids in beer. And, of course, the end products of some fermentations are historically the basis of protection for fermented foodstuffs, for example, low pH, organic acids, alcohol, carbon dioxide. Of especial interest here is nisin (Fig. 1.9) that is a natural product from lactic acid bacteria, capable of countering the invasion of other bacteria.

An essential aspect of the long-term success of lactic acid bacteria as a protective agent within the fermentation industries is the multiplicity of ways in which it counters the growth of competing organisms. Apart from nisin and other bacteriocins, we might draw attention to the production of

(1) organic acids, such as lactic, acetic and propionic acids, with acetic acid being especially valuable in countering bacteria, yeasts and moulds;
(2) hydrogen peroxide, which, as we have seen is an activated (and therefore potentially damaging) derivative of oxygen;
(3) diacetyl and acetaldehyde, although some argue that the levels developed are not of practical significance as antimicrobial agents.

Table 1.4   Food grade antimicrobial agents.

*Preservative*
Acetic acid and its sodium, potassium and calcium salts
Benzoic acid and its sodium, potassium and calcium salts
Biphenyl
Formic acid and its sodium and calcium salts
Hydrogen peroxide
$p$-Hydroxybenzoate, ethyl-, methyl- and propyl variants and
  their sodium salts
Lactic acid
Nisin
Nitrate and nitrite, and its sodium and potassium salts
$o$-Phenylphenol
Propionic acid and its sodium, potassium and calcium salts
Sorbic acid and its sodium, potassium and calcium salts
Sulphur dioxide, sodium and potassium sulphites, sodium and
  potassium bisulphites, sodium and potassium metabisulphites
  (disulphites)
Thiabendazole

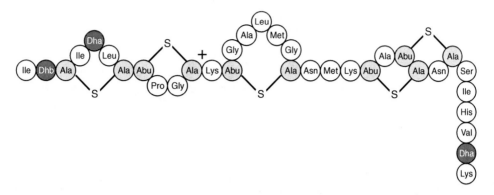

**Fig. 1.9** *Nisin*. This antimicrobial destroys Gram-positive organisms by making pores in their membranes. It includes some unusual amino acids, including dehydrated serine (Dha), dehydrated threonine (Dhb), lanthionine (Ala–S–Ala) and $\beta$-methyllanthionine (Abu–S–Ala). The last two originate from the coupling of cysteine with dehydrated serine or threonine, respectively. See also http://131.211.152.52/research_page/nisin.html.

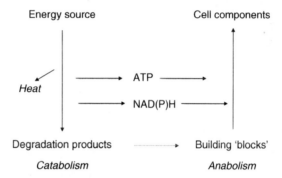

**Fig. 1.10** Energy sources (e.g. sugars) are successively broken down in catabolic reactions, resulting in the capture of energy in the form of ATP and reducing power (as reduced NADH). Building blocks are transformed into the polymers from which cells are comprised (see Figs 1.3–1.6) in anabolic reactions that draw on energy (ATP) and reducing power (many of the anabolic processes use the phosphorylated form of NADH, i.e. NADPH).

## Metabolic events

### Catabolism

Catabolism refers to the metabolic events whereby a foodstuff is broken down so as to extract energy in the form of adenosine triphosphate (ATP), as well as reducing power (customarily generated primarily in the form of nicotinamide adenine dinucleotide (NADH, reduced form) but utilised as nicotinamide adenine dinucleotide phosphate (NADPH, reduced form) to fuel the reactions (anabolism) wherein cellular constituents are fabricated (Fig. 1.10).

In focusing on the organotrophs, and in turn even more narrowly (for the most part) on those that use sugars as the main source of carbon and energy, we must first consider the Embden–Meyerhof–Parnas (EMP)

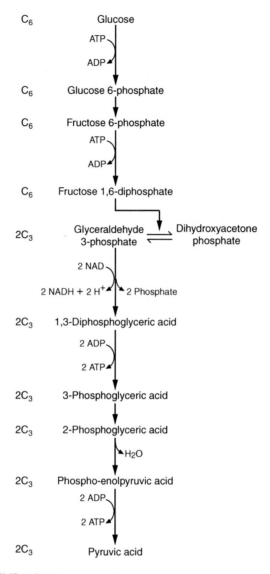

$C_6$    Glucose

ATP

ADP

$C_6$    Glucose 6-phosphate

$C_6$    Fructose 6-phosphate

ATP

ADP

$C_6$    Fructose 1,6-diphosphate

$2C_3$    Glyceraldehyde 3-phosphate ⇌ Dihydroxyacetone phosphate

2 NAD

2 NADH + 2 H$^+$    2 Phosphate

$2C_3$    1,3-Diphosphoglyceric acid

2 ADP

2 ATP

$2C_3$    3-Phosphoglyceric acid

$2C_3$    2-Phosphoglyceric acid

$H_2O$

$2C_3$    Phospho-enolpyruvic acid

2 ADP

2 ATP

$2C_3$    Pyruvic acid

**Fig. 1.11**    The EMP pathway.

pathway (Fig. 1.11). This is the most common route by which sugars are converted into a key component of cellular metabolism, pyruvic acid. This pathway, for example, is central to the route by which alcoholic fermentations are performed by yeast. In this pathway, the sugar is 'activated' to a more reactive phosphorylated state by the addition of two phosphates from ATP. There follows a splitting of the diphosphate to two three-carbon units that are in equilibrium. It is the glyceraldehyde 3-phosphate that is metabolised further, but as it is used up, the equilibrium is strained and dihydroxyace-tone phosphate is converted to it. Hence we are in reality dealing with two identical units proceeding from the fructose diphosphate. The first step is oxidation, the reducing equivalents (electrons, hydrogen) being captured by NAD. En route to pyruvate are two stages at which ATP is produced by the splitting off of phosphate – this is called *substrate-level phosphorylation*. As there are two three-carbon ($C_3$) fragments moving down the pathway, this therefore means that four ATPs are being produced per sugar molecule. As two ATPs were consumed in activating the sugar, there is a net ATP gain of two.

In certain fermentations, the Entner–Doudoroff pathway (Fig. 1.12) is employed by the organism, a pathway differing in the earliest part insofar as only one ATP is used. Meanwhile, in certain lactic acid bacteria, there is the quite different phosphoketolase pathway (Fig. 1.13).

A major outlet for pyruvate is into the Krebs cycle (tricarboxylic acid cycle; Fig. 1.14). In particular, this cycle is important in aerobically growing cells. There are four oxidative stages with hydrogen collected either by NAD or flavin adenine dinucleotide (FAD). When growing aerobically, this reducing power can be recovered by successively passing the electrons across a sequence of cytochromes located in the mitochondrial membranes of eukaryotes or the plasma membrane of prokaryotes (Fig. 1.7), with the resultant flux of protons being converted into energy collection as ATP through the process of oxidative phosphorylation (Fig. 1.15). In aerobic systems, the terminal electron acceptor is oxygen, but other agents such as sulphate or nitrate can serve the function in certain types of organism. An example of the latter would be the nitrate reducers that have relevance in certain meat fermentation processes (see Chapter 13).

In classic fermentations where oxygen is not employed as a terminal electron acceptor and indeed the respiratory chain as a whole is not used, there needs to be an alternative way for the cell to recycle the NADH produced in the EMP pathway, so that NAD is available to continue the process. Herein lies the basis of much of the diversity in fermentation end products, with pyruvate being converted in various ways (Fig. 1.16). In brewer's yeast, the end product is ethanol. In lactic acid bacteria, there are two modes of metabolism. In *homofermentative* bacteria, the pyruvate is reduced solely to lactic acid. In *heterofermentative* lactic acid bacteria, there are alternative end products, most notably lactate, ethanol and carbon dioxide, produced through the intermediacy of the phosphoketolase pathway.

As noted earlier, higher molecular weight molecules that are too large to enter into the cell *as is* are hydrolysed by enzymes secreted from the organism. The resultant lower molecular weight materials are then transported into the cell in the same manner as exiting smaller sized materials. The transport is by selective permeases, which are elaborated in response to the needs of the cell. For example, if brewing yeast is exposed to a mixture of sugars, then it will elaborate the transport permeases (proteins) in a defined sequence (see Chapter 2).

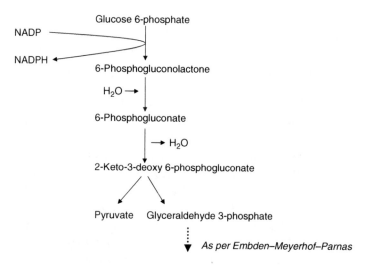

**Fig. 1.12**    The Entner–Doudoroff pathway.

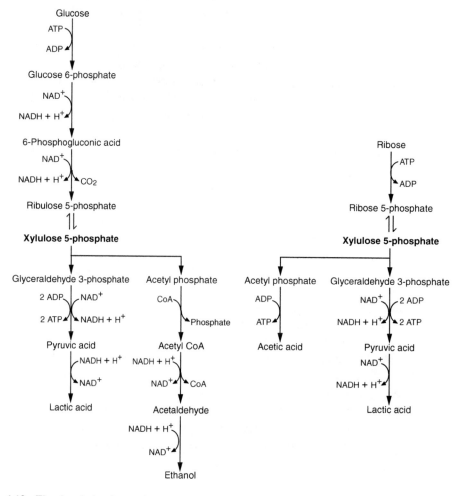

**Fig. 1.13**    The phosphoketolase pathway.

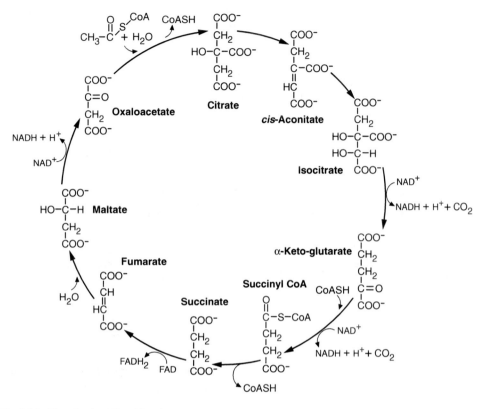

**Fig. 1.14**  The tricarboxylic acid cycle.

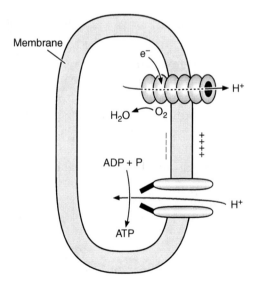

**Fig. 1.15**  Oxidative phosphorylation. The passage of electrons through the electron transport chain is accompanied by an exclusion of protons ($H^+$) from the cell (or mitochondrion for a eukaryote). The energetically favourable return passage of protons 'down' a concentration gradient is linked to the phosphorylation of ADP to produce ATP.

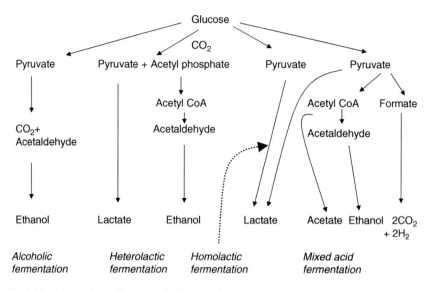

**Fig. 1.16**   Alternative end products in fermentation.

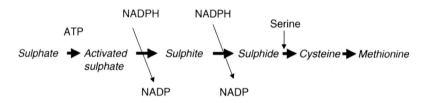

**Fig. 1.17**   The assimilation of sulphur.

*Anabolism*

The above-named pathways are examples of how cells deal with sugars, thereby obtaining carbon, hydrogen and oxygen. As observed earlier, cells must also secure a supply of other elements from the medium. Nitrogen may be provided as amino acids (e.g. in the case of brewing yeast), urea or inorganic nitrogen forms, primarily as ammonium salts (often used in wine fermentations).

Sulphur can variously be supplied in organic or inorganic forms. Brewing yeast, for example, can assimilate sulphate, but will also take up sulphur-containing amino acids (Fig. 1.17).

The major structural and functional molecules in cells are polymeric. These include

(1) Polysaccharides – notably the storage molecules such as glycogen in yeast, which has a structure closely similar to the amylopectin fraction of starch (see later), and the structural components of cell walls, for example, the mannans and glucans in yeast and the complex polysaccharides in bacterial cell walls.

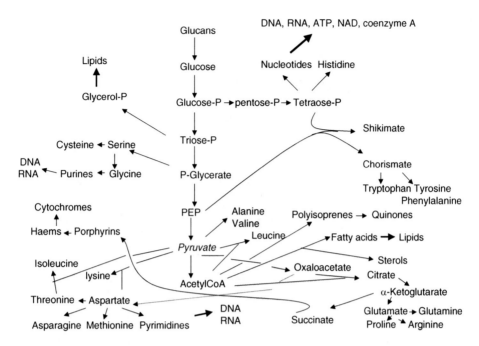

**Fig. 1.18**  A simplified overview of intermediary metabolism.

(2)  Proteins – notably the enzymes and the permeases.
(3)  Lipids – notably the components at the heart of membrane structure.
(4)  Nucleic acids – DNA and RNA.

A greatly simplified summary of cellular metabolism, incorporating the essential features of anabolic reactions is given in Fig. 1.18. It is sufficient in the present discussion to state that pyruvate is at the heart of the metabolic pathways. There are clearly various draws on it, both catabolic and anabolic. Of particular note is the draw off from the tricarboxylic acid cycle to satisfy biosynthetic needs, meaning that there is a failure to regenerate the oxaloacetate needed to collect a new acetyl-CoA residue emerging from pyruvate. Thus, cells have so-called anaplerotic pathways by which they can replenish necessary intermediates such as oxaloacetate. The best-known such pathway is the glyoxylate cycle (Fig. 1.19).

It is essential that the multiplicity of reactions, which as a whole constitute cellular metabolism, are controlled so that the whole is in balance to achieve the appropriate needs of the cell under the prevailing conditions within which it finds itself. It is outside the scope of this book to dwell on these regulatory mechanisms, but they include coarse controls on the synthesis of the necessary permeases and enzymes (the general rule being that a protein is only synthesised as and when it is needed) and fine controls on the rate at which the enzymes are able to act. Examples of the impact of these control strategies

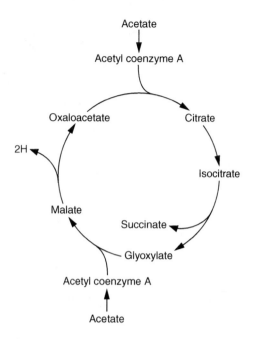

**Fig. 1.19**  The glyoxylate cycle.

will be encountered in this book, for example, whether brewing yeast degrades sugars by respiration or fermentation.

## The origins of the organisms employed in food fermentations

For the longest time, the foodstuffs described in this book were prepared using endogenous microflora. Increasingly, however, and starting first with the isolation of pure strains of brewing yeast by Emil Christian Hansen in 1883, many of the products employ starter cultures in their production. The organisms conform to the criterion of being Generally Recognised as Safe (GRAS). They are selected for their advantageous properties in terms of process performance and impact on final product quality.

Many companies and academic laboratories are seeking newer, improved cultures. This can be achieved in what may be called 'serendipity mode' by screening a broad swathe of samples taken from multitudinous habitats, the screening employing growth media and cultivation conditions that are best suited to an organism with the desired characteristics. Alternatively, some narrowing of odds can be achieved by specifically looking in locales where certain types of organisms are known to thrive – for example, yeasts are plentiful on the surface of fruit. One extreme example of this approach might most reasonably be described as 'theft', with the pure culture of one company finding its way, through whatever mechanism, into the clutches of another corporation.

**Table 1.5**  Culture collections.

| Collection | Organisms | Web page |
|---|---|---|
| American Type Culture Collection (ATCC) | All types | http://www.atcc.org/ |
| CABI Bioscience | Filamentous fungi | http://www.cabi-bioscience.org/ |
| Centraalbureau voor Schimmelcultures | Filamentous fungi and yeasts | http://www.cbs.knaw.nl/ |
| Collection Nationale de Cultures de Microorganismes | All types | http://www.pasteur.fr/recherche/unites/Cncm/index-en.html |
| Die Deutsche Sammlung von Mikroorganismen und Zellkulturen | All types | http://www.dsmz.de/ |
| Herman J. Phaff Culture Collection | Yeasts and fungi | http://www.phaffcollection.org/ |
| National Collection of Industrial and Marine Bacteria | Bacteria | http://www.ncimb.co.uk/ |
| National Collection of Yeast Cultures | Yeasts | http://www.ifr.bbsrc.ac.uk/NCYC/ |

A more honest approach is by purchasing samples of pure organisms of the desired character from culture collections (Table 1.5). Nowadays the cultures are likely to be in the form of vials frozen in liquid nitrogen ($-196°C$) or they may be lyophilised. For some industries, notably bread making and wine making, companies do not produce their own yeast but rather bring it into the production facility on a regular basis from a supplier company. This might be supplied frozen or merely refrigerated with cryoprotectants such as sucrose, glycerol or trehalose. The latest technology here is active dried yeast, with the organism cultured optimally to ensure its ability to survive drying in a state that will allow it to perform vigorously and representatively when re-hydrated. In other industries, notably beer brewing, companies tend to maintain their own strains of yeast and propagate these themselves. This is probably on account of the fact that beer-making is essentially the only industry described in this book where the surplus organism that grows in the process is re-used.

An overview of starter cultures is given in Table 1.6. A starting inoculum might typically be of the order of 1%. An example of how the volume can be scaled up from the pure 'slope' of the master culture to an amount to 'pitch' the most enormous of fermenters is given in Chapter 2.

There are various opportunities for enhancing the properties of the organisms that are already employed in food companies. Mutagenesis to eliminate undesirable traits has been employed. However, this is a challenge for eukaryotes as such cells tend to have multiple copies of each gene (polyploidy), and it is a formidable challenge to eliminate all the alleles of the undesirable gene. Classic recombination techniques (conjugation, transduction and transformation) have been pursued, but there is always the risk that an undesirable trait

**Table 1.6**  Starter cultures.

| Organism | Type of organism | Foodstuff |
| --- | --- | --- |
| *Aspergillus oryzae* | Mould | Miso, soy sauce |
| *Brevibacterium linens* | Bacterium | Cheese pigment and surface growth |
| *Lactobacillus casei* | Bacterium | Cheese and other fermented dairy products |
| *Lactobacillus curvatus* | Bacterium | Sausage |
| *Lactobacillus delbrueckii* ssp. *bulgaricus* | Bacterium | Cheese, yoghurt |
| *Lactobacillus helveticus* | Bacterium | Cheese and other fermented dairy products |
| *Lactobacillus lactis* (various ssp.) | Bacterium | Cheese and other fermented dairy products |
| *Lactobacillus plantarum* | Bacterium | Fermented vegetables, sausage |
| *Lactobacillus sakei* | Bacterium | Sausage |
| *Lactobacillus sanfranciscensis* | Bacterium | Sourdough bread |
| *Leuconostoc lactis* | Bacterium | Cheese and other fermented dairy products |
| *Leuconostoc mesenteroides* | Bacterium | Fermented vegetables, cheese and other fermented dairy products |
| *Oenococcus oeni* | Bacterium | Wine |
| *Pediococcus acidilactici* | Bacterium | Fermented vegetables, sausage |
| *Pediococcus halophilus* | Bacterium | Soy sauce |
| *Pediococcus pentosaceus* | Bacterium | Sausage |
| *Penicillium camemberti* | Mould | Surface ripening of cheese |
| *Penicillium chrysogenum* | Mould | sausage |
| *Penicillium roqueforti* | Mould | Blue-veined cheeses |
| *Propionibacterium freudenreichii* | Bacterium | Eyes in Swiss cheese |
| *Rhizopus microsporus* | Mould | Tempeh |
| *Saccharomyces cerevisiae* | Fungus | Bread, ale, wine |
| *Saccharomyces pastorianus* | Fungus | Lager |
| *Staphylococcus carnosus* | Fungus | Meat |
| *Streptococcus thermophilus* | Bacterium | Cheese, yoghurt |

will be introduced as an accompaniment to the trait of interest. Much more selectivity is afforded by modern genetic modification strategies. However, as noted earlier, this attracts far more emotion for organisms used in food production than it does in the production of, say, fuels or pharmaceuticals.

## Some of the major micro-organisms in this book

Reference to the chapters that follow will highlight to the reader that a diversity of micro-organisms is involved in food fermentations. However, the organisms that one encounters most widely in these processes are undoubtedly the yeasts, notably Saccharomyces, and lactic acid bacteria. It is important to note in passing that if these organisms 'stray' from where they are supposed to be, then they are spoilage organisms with a ruinous nature. For example, lactic acid bacteria have a multiplicity of values in the production of many foodstuffs, including cheese, sourdough bread, some wines and a very few beers. However, their development in the majority of beers is very much the primary source of spoilage.

## *Yeast*

In most instances, use of the word yeast in a food context is synonymous with *S. cerevisiae*, namely, brewer's yeast or baker's yeast. However, as we shall discover, there are other yeasts involved in fermentation processes.

Yeasts are heterotrophic organisms whose natural habitats are the surfaces of plant tissues, including flowers and fruit. They are mostly obligate aerobes, although some (such as brewing yeast) are facultative anaerobes. They are fairly simple in their nutritional demands, requiring a reduced carbon source, various minerals and a supply of nitrogen and vitamins. Ammonium salts are readily used, but equally a range of organic nitrogen compounds, notably the amino acids and urea, can be used. The key vitamin requirements are biotin, pantothenic acid and thiamine.

Focusing on brewing yeast, and following the most recent taxonomic findings, the term *S. cerevisiae* is properly applied only to ale yeasts. Lager yeasts are properly termed *Saccharomyces pastorianus,* representing as they do organisms with a 50% larger genome and tracing their pedigree to a coupling of *S. cerevisiae* with *Saccharomyces bayanus*.

Saccharomyces (see Fig. 1.2) is spherical or ellipsoidal. Whereas laboratory strains are haploid (one copy of each of the 16 linear chromosomes), industrial strains are polyploid (i.e. they have multiple copies of each chromosome) or aneuploid (varying numbers of each chromosome). Some 6000 genes have been identified in yeast and indeed the entire genome has now been sequenced (see http://www.yeastgenome.org/).

Brewing yeast does have a sex life, but reproduces in production conditions primarily by budding (Fig. 1.20). A single cell may bud up to 20 times, each time leaving a scar, the counting of which indicating how senile the cell has become.

The surface of the wall surrounding the yeast cell is negatively charged due to the presence of phosphate groups attached to the mannan polysaccharides that are located in the wall. This impacts the extent to which adjacent cells can interact, and the presence of calcium ions serves to bridge cells through ionic bonding. Coupled with other interactions between lectins in the surface, there are varying degrees of association between different strains, resulting in differing extents of flocculation, advantage of which is taken in the separation of cells from the liquid at the end of fermentation.

The underlying plasma membrane (as well as the other membranes in the cellular organelles) is comprised primarily of sterols (notably ergosterol), unsaturated fatty acids and proteins, notably the permeases (discussed earlier) (Fig. 1.21). As oxygen is needed for the desaturation reactions involved in the

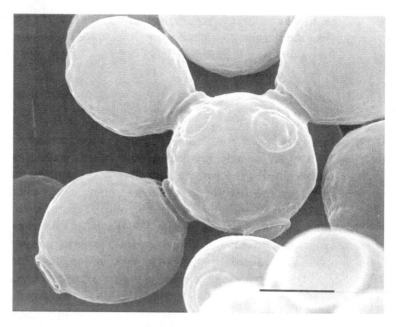

**Fig. 1.20**   Yeast cells budding. Bud scars, where previous cell division has occurred, are visible. Photograph courtesy of Dr Alastair Pringle.

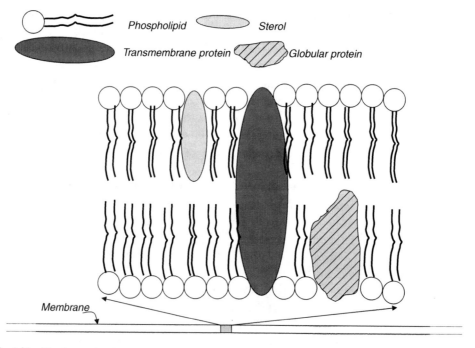

**Fig. 1.21**   Membrane structure.

synthesis of the lipids, relatively small quantities of oxygen must be supplied to the yeast, even when it is growing anaerobically by fermentation.

The control mechanisms that drive the mode of metabolism in the yeast cell (i.e. by aerobic respiration or by fermentation) are based on the concentration of sugar that the yeast is exposed to. At high concentrations of sugar, the cell is switched into the fermentative mode, and the pyruvate is metabolised via acetaldehyde to ethanol. At low sugar concentrations, the pyruvate shunts into acetyl-CoA and the respiratory chain. This is the so-called Crabtree effect. The rationale is that when sugar concentrations are high, the cell does not need to generate as many ATP molecules per sugar molecule, whereas if the sugar supply is limited, the yeast must maximise the efficiency with which it utilises that sugar (ATP yield via fermentation and respiration are 2 molecules and 32 molecules, respectively). The significance of this in commercial fermentation processes is clear. In brewing, where the primary requirement is a high yield of alcohol, the sugar content in the feedstock (wort) is high, whereas in the production of baker's yeast, where the requirement is a high cell yield, the sugar concentration is always kept low, but the sugar is continuously passed into the fermenter ('fed batch').

## Lactic acid bacteria

Throughout the centuries it has been the practice in various fermentation-based processes to add back a proportion of the previously produced food to the new batch, so-called back slopping. What of course this did was to seed the fermentation with the preferred micro-organism, and for many foodstuffs this organism is a lactic acid bacterium. Such bacteria are only weakly proteolytic and lipolytic, which means that they are quite 'mild' with respect to their tendency to produce pungent flavours. They are also naturally present in the intestine and the reproductive tract, so it is no surprise that nowadays we talk of probiotics and prebiotics in the context of enriching the level of lactic acid bacteria in the gut. Probiotics are organisms, notably lactobacilli and bifidobacteria, which are added to the diet to boost the flora in the large intestine. For example, they are added to yoghurt. Prebiotics are nutrients that boost the growth of these organisms.

Like the brewing and baking yeasts, lactic acid bacteria tend to be GRAS, although some strains are pathogenic. Joseph Lister isolated the first lactic acid bacterium in 1873. This organism that we now refer to as *Lactococcus lactis* is a species of great significance in the fermentation of milk products.

There are 16 genera of lactic acid bacteria, some 12 of which are active in a food context. They are Gram-positive organisms, are either rod-shaped, cocci (spherical) or coccobacilli. For the most part they are mesophilic, but some can grow at refrigerator temperatures (4°C) and as high as 45°C. Generally they prefer a pH in the range 4.0–4.5, but certain strains can tolerate and grow at pHs above 9.0 or as low as 3.2. They need preformed purines, pyrimidines, amino acids and B vitamins. Lactic acid bacteria do not possess a functional tricarboxylic acid cycle or haem-linked electron transport systems, so they use substrate level phosphorylation to gain their energy.

As we saw previously, their metabolism can be classified as either homofermentative, where lactic acid represents 95% of the total end products, or heterofermentative, in which acetic acid, ethanol and carbon dioxide are produced alongside lactic acid.

Lactic acid bacteria produce antimicrobial substances known as bacteriocins. For the most part, these are cationic amphipathic peptides that insert into the membranes of closely related bacteria, causing pore formation, leakage and an inability to sustain metabolism, ergo death. The best known of these agents is nisin (discussed earlier), which has been used substantially as a 'natural' antimicrobial agent. Lactic acid bacteria also produce acids and hydrogen peroxide as antimicrobials.

### Lactococcus

The most notable species within this genus is *L. lactis*, which is most important in the production of foodstuffs such as yoghurts and cheese. It is often co-cultured with Leuconostoc.

There are two sub-species of *L. lactis*: Cremoris, which is highly prized for the flavour it affords to certain cheese, and Lactis, in particular *L. lactis* ssp. *lactis* biovar. *diacetyllactis*, which can convert citrate to diacetyl, a compound with a strong buttery flavour highly prized in some dairy products but definitely taboo in most, if not all, beers. The carbon dioxide produced by this organism is important for eye formation in Gouda.

### Leuconostoc

These are heterofermentative cocci.

*Leuconostoc mesenteroides,* with its three subspecies: mesenteroides, cremoris and dextranicum, and *Leuc. lactis* are the most important species, especially in the fermentation of vegetables. They produce extracellular polysaccharides that have value as food thickeners and stabilisers. These organisms also contribute to the $CO_2$ production in Gouda.

*Oenococcus oeni* (formerly *Leuc. oenos*) plays an important role in malolactic fermentations in wine.

### Streptococcus

These are mostly pathogens; however, *Streptococcus thermophilus* is a food organism, featuring alongside *Lactobacillus delbrueckii* ssp. *bulgaricus* in the production of yoghurt. Furthermore, it is used in starter cultures for certain cheeses, notably Parmesan.

*Lactobacillus*

There are some 60 species of such rod-shaped bacteria that inhabit the mucous membranes of the human, *ergo* the oral cavity, the intestines and the vagina. However, they are equally plentiful in foodstuffs, such as plants, meats and milk products.

*Lb. delbrueckii* ssp. *bulgaricus* is a key starter organism for yoghurts and some cheeses. However, lactobacilli have involvement in other fermentations, such as sourdough and fermented sausages, for example, salami. Conversely, they can spoil beer and either fresh or cooked meats, etc.

*Pediococci*

*Pediococcus halophilus* (now *Tetragenococcus*) is extremely tolerant of salt (>18%) and as such is important in the production of soy sauce. Pediococci also function in the fermentation of vegetables, meat and fish. On the other hand, *Pediococcus damnosus* growth results in ropiness in beer and the production of diacetyl as an off-flavour.

*Enterococcus*

These faecal organisms have been isolated from various indigenous fermented foods; however, no positive contribution has been unequivocally demonstrated and their presence is debatably indicative of poor hygiene.

## Providing the growth medium for the organisms

The microflora is of course one of the two key inputs to food fermentation. The other is the substrate that the organism(s) converts. With the possible exception of mycoprotein (see Chapter 17), the substrates that we encounter in this book are very traditional and well-defined insofar as the end product is what it is as much because of that substrate as through the action of the microbe that deals with it. Thus, for beers, the final product, whether it is an ale, lager or stout, a wheat beer or a lambic has clear characteristics that are afforded by the raw materials (malt, adjunct and hops) used to make the wort that the yeast ferments. The same applies for the cereal used to make bread, the milk going to cheese and yoghurt, the meat destined for salami, the cabbage en route to sauerkraut.

In all instances there are defined preparatory steps that must be undertaken to render the substrate in the state that is ready for the microbial fermentative activity. For some foodstuffs (e.g. yoghurt), there is relatively little processing of the milk. However, for a product like beer, there is prolonged initial processing, notably the malting of grain and its subsequent extraction in the brewery.

The growth substrate must always include sources of carbon, nitrogen, water and, usually, oxygen, as well as the trace elements. These nutritional considerations have already been discussed.

## Fermenters

Most food fermentations are generally classified as being 'non-aseptic' to distinguish them from microbial processes where rigorous hygiene must be ensured, for example, production of antibiotics and vaccines. This is not to say that those practising food fermentations are less than hygienic. The majority of the processes that I describe in this book are carried out in vessels that are subject to rigorous CIP (discussed earlier).

A diversity of fermenter types is employed ranging from the relatively sophisticated cylindroconical vessels in modern brewery operations (see Fig. 2.25) through to the relatively crude set-ups used in some of the indigenous fermentation operations, not the least the fermentation of cocoa. Key issues in all instances are the ability to maintain the required degree of cleanliness, the ability to mix, the ability to regulate temperature and change temperature smoothly and efficiently, the access of oxygen (aeration or oxygenation) and the ability to monitor and control.

## Downstream processing

For many of the foodstuffs that we address, some form of post-fermentation clarification is necessary to remove surplus microbial cells and various other types of insoluble particles. Cells may be harvested by sedimentation (perhaps encouraged by agents such as isinglass or egg white), centrifugation or filtration. Additionally, there may be other downstream treatments, such as the adsorption of materials that might (if not removed) fall out of solution and ruin the appearance of a product, for example, polyphenols and proteins in beer. Many products have their microbial populations depleted either by pasteurisation or filtration through depth and/or membrane filters. Finally, of course, they receive varying degrees of primary and secondary packaging.

Several of the products described in the present volume involve distillation stage(s) in their production. This will be described in general terms in Chapter 6.

## Some general issues for a number of foodstuffs

Some topics are of general significance for many of the foodstuffs considered in this book and, accordingly, reference is made to them here.

### Non-enzymatic browning

These are chemical reactions that lead to a brown colour when food is heated. The relevant chemistry is known as the Maillard reaction, which actually comprises a sequence of reactions that occurs when reducing sugars are heated with compounds that contain a free amino group, for example, amino acids, proteins and amines (Fig. 1.22, Table 1.7). In reflection of the complexity of the chemistry, there are many reaction intermediates and products. As well as colour, Maillard reaction products have an impact on flavour and may act as antioxidants. These antioxidants are mostly produced at higher pHs and when the ratio of amino acid to sugar is high. It must also be stressed that some of the Maillard reaction products can *promote* oxidative reactions. Other Maillard-type reactions occur between amino compounds and substances other than sugars that have a free carbonyl group. These include ascorbic acid and molecules produced during the oxidation of lipids.

The Maillard reaction should not be confused with caramelisation, which is the discoloration of sugars as a result of heating in the absence of amino compounds.

In the primary Maillard reaction, the amino compound reacts with the reducing sugar to form an N-substituted glycosylamine that rearranges to 1-amino-1-deoxy-2-ketose (the so-called Amadori rearrangement product). This goes forward in a cascade of reactions in various ways depending on the pH. At the pH of most foods (4–6), the primary route involves melanoidin formation by further reaction with amino acids. Other products are Strecker aldehydes, pyrazines, pyrolles and furfurals. The substances produced in these reactions have flavours that are typical of roasted coffee and nuts, bread and cereals. The pyrolle derivatives afford bitter tastes. The Maillard reaction may

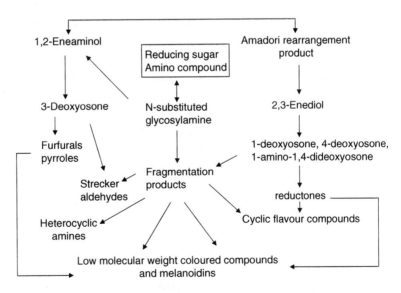

**Fig. 1.22**   The Maillard reaction.

**Table 1.7**  Some products of the Maillard reaction.

| Type of compound | Example | Flavour descriptors |
|---|---|---|
| *Products derived from interactions of sugars and amino acids* | | |
| Pyrolle | 2-Acetyl-1-pyrroline | Newly baked crust of wheat bread |
| Pyridine | 2-Acetyl-1,4,5,6-tetrahydropyridine | Cream crackers |
| Pyrazine | Methylpyrazine | Nut |
| Oxazole | Trimethyloxazole | Green, nutty, sweet |
| Thiophene | 2-Acetylthiphene | Onion, mustard |
| *Products derived from the sugar* | | |
| Furan | Furaneol | Caramel, strawberry |
| Carbonyl | Diacetyl | Butterscotch |
| *Products derived from the amino acid* | | |
| Cyclic polysulphur | 5-Methyl-5-pentyl-1,2,4-trithiolane | Fried chicken |
| Sulphur-container | Methional | Mashed potato |
| Thiazole | 2-Acetylthiazole | Popcorn |

also lead to aged or cooked characters in products such as processed orange juice and dried milk products.

The early products in the Maillard reaction are colourless, but when they get progressively larger, they become coloured and responsible for the hue of a wide range of foods. Some of these coloured compounds have low molecular weights, but others are much larger and may include complexes produced by heat-induced reactions of the smaller compounds and proteins.

The exact events in any Maillard-based process depend on the proportion of the various precursors, the temperature, pH, water activity and time available. Metals, oxygen and inhibitors such as sulphite also impact. The flavour developed differs depending on the time and intensity of heating for instance – high temperature for a short time gives a different result when compared with low temperature for a long time. Pentose sugars react faster than do hexoses, which in turn react more rapidly than disaccharides such as maltose and lactose. With regard to the amino compounds, lysine and glycine are much more reactive than is cysteine, for instance, but more than that, for the flavour also depends on the amino acid. Cysteine affords meaty character; methionine gives potato, while proline gives bready.

As water is produced in the Maillard reaction, it occurs less readily in foods where the water activity is high. The Maillard reaction is especially favoured at $A_w$ 0.5–0.8.

Finally sulphite, by combining with reducing sugars and other carbonyl compounds, inhibits the reaction.

## *Enzymatic browning*

This arises by the oxidation of polyphenols to *o*-quinones by enzymes such as polyphenol oxidase (PPO) and peroxidase (Fig. 1.23). A day-to-day example

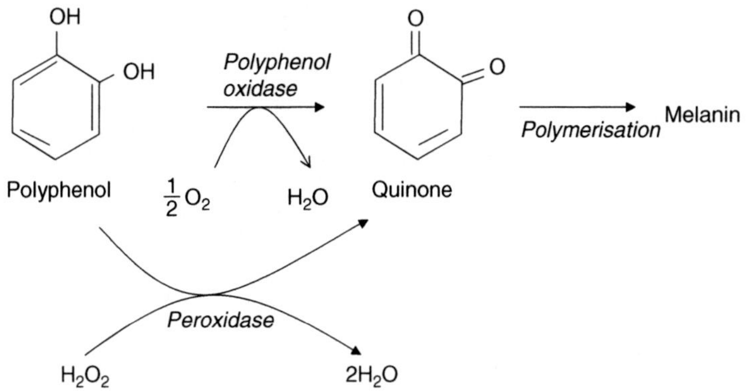

**Fig. 1.23** Polyphenol oxidation.

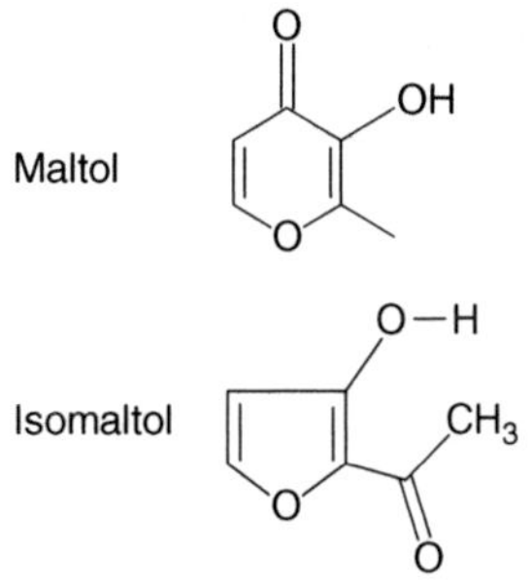

**Fig. 1.24** Some flavour compounds produced in caramelisation reactions.

would be the browning of sliced apple. In other foods, the reaction is wanted, for example, in the readying of prunes, dates and tea for the marketplace.

Whereas heating boosts non-enzymatic browning, the converse applies to enzymatic browning, as the heat inactivates enzymes. The alternative strategies to avoid the reaction are to lower the levels of polyphenols (the agent polyvinylpolypyrrolidone (PVPP) achieves this) or to exclude oxygen.

## Caramel

This is still produced to this day by burning sugar, but in very controlled ways. The principal products are produced by the polymerisation of glucose by dehydration. The process is catalysed by acids or bases and requires temperatures in excess of 120°C. In some markets, the word caramel is retained for materials that are produced in the absence of nitrogen-containing compounds and these products are used for flavouring value. Where N is present, then 'sugar colours' are produced and these are used for colouring purposes.

Caramel is polymeric in nature, but also contains several volatile and non-volatile lower molecular weight components that afford the characteristic flavour compounds, such as maltol and isomaltol (Fig. 1.24).

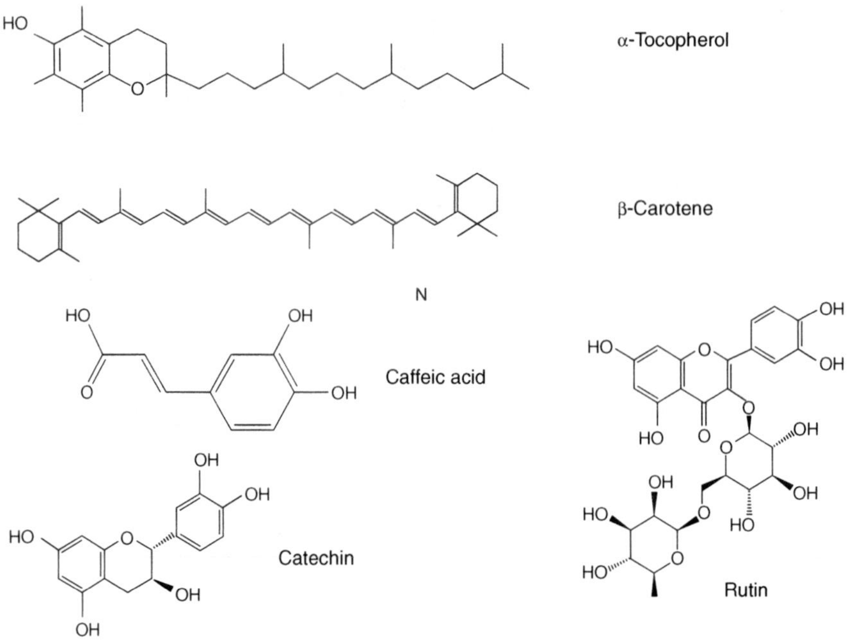

**Fig. 1.25**  Some antioxidants.

## *Antioxidants*

There is much interest in antioxidants from the perspective of protecting foodstuffs from flavour decay, but increasingly for their potential value in countering afflictions such as cancer, rheumatoid arthritis and inflammatory bowel diseases. Figure 1.25 presents a range of these antioxidants. Many are phenolics and act either by scavenging or by neutralising (reduction) the radicals that effect deterioration or by chelating the metal ions that cause the production of these radicals.

The tocopherols are fat soluble and are found in vegetable oils and the fatty regions of cereals, for example, the germ. The carotenoids (e.g. lycopene) are water soluble and are found in fruits and vegetables. The flavonoids are water-soluble polyphenols found in fruits, vegetables, leaves and flowers. Such molecules have particular significance for some of the products discussed in this book, notably wine, beer and tea. The phenolic acids, for example, caffeic and ferulic acids and their esters, are abundant in cereal grains such as wheat and barley.

## Bibliography

Anke, T. (1997) *Fungal Biotechnology*. London: Chapman & Hall.

Atkinson, B. & Mavituna, F. (1991) *Biochemical Engineering and Biotechnology Handbook*, 2nd edn. Basingstoke: Macmillan.

Berry, D.R., ed. (1988) *Physiology of Industrial Fungi.* Oxford: Blackwell.

Branen, A.L. & Davidson, P.M., eds (1983) *Antimicrobials in Foods.* New York: Marcel Dekker.

Brown, C.M., Campbell, I. & Priest, F.G. (1987) *Introduction to Biotechnology.* Oxford: Blackwell Publishing.

Caldwell, D.R. (1995) *Microbial Physiology and Metabolism.* Oxford: William C. Brown.

Dawes, I.W. & Sutherland, I.W. (1992) *Microbial Physiology*, 2nd edition. Oxford: Blackwell Publishing.

Demain, A.L., Davies, J.E. & Atlas, R.M. (1999) *Manual of Industrial Microbiology and Biotechnology.* Washington, DC: American Society for Microbiology.

Frankel, E.N. (1998) *Lipid Oxidation.* Dundee: Oily Press.

Griffin, D.H. (1994) *Fungal Physiology*, 2nd edn. New York: Wiley-Liss.

Jennings, D.M. (1995) *The Physiology of Fungal Nutrition.* Cambridge: Cambridge University Press.

Lengeler, J.W., Drews, G. & Schlegel, H.G. (1999) *Biology of the Prokaryotes.* Oxford: Blackwell Publishing.

McNeil, B. & Harvey, L.M. (1990) *Fermentation: A Practical Approach.* Oxford: IRL.

O'Brien, J., Nursten, H.E., Crabbe, M.J.C. & Ames, J.M., eds (1998) *Maillard Reaction in Foods and Medicine.* Cambridge: Royal Society of Chemistry.

Pirt, S.J. (1975) *The Principles of Microbe and Cell Cultivation.* Oxford: Blackwell Publishing.

Salminen, S. & Von Wright, A., eds (1998) *Lactic Acid Bacteria: Microbiology and Functional Aspects.* New York: Marcel Dekker.

Stanbury, P.F., Whitaker, A. & Hall, S.J. (1995) *Principles of Fermentation Technology*, 2nd edn. Oxford: Butterworth-Heinemann (Pergamon).

Tucker, G.A. & Woods, L.F.J. (1995) *Enzymes in Food Processing.* London: Blackie.

Waites, M.J., Morgan, N.L., Rockey, J.S. & Higton, G. (2001). *Industrial Microbiology: An Introduction.* Oxford: Blackwell Publishing.

Walker, G.M. (1998) *Yeast Physiology and Biotechnology.* Chichester: Wiley.

Ward, O.P. (1989) *Fermentation Biotechnology: Principles, Processes and Products.* UK: Open University Press.

Wood, B.J.B. and Holzapfel, W.H. (1996) *The Genera of Lactic Acid Bacteria.* London: Blackie.

# Chapter 2
# Beer

The word beer comes from the Latin word *Bibere* (to drink). It is a beverage whose history can be traced back to between 6000 and 8000 years and the process, being increasingly regulated and well controlled because of tremendous strides in the understanding of it, has remained unchanged for hundreds of years. The basic ingredients for most beers are malted barley, water, hops and yeast; indeed, the 500-year-old Bavarian purity law (the *Reinheitsgebot*) restricts brewers to these ingredients for beer to be brewed in Germany. Most other brewers worldwide have much greater flexibility in their production process opportunities, yet the largest companies are ever mindful of the importance of tradition.

Compared to most other alcoholic beverages, beer is relatively low in alcohol. The highest average strength of beer (alcohol by volume (ABV) indicates the millilitres of ethanol per 100 ml of beer) in any country worldwide is 5.1% and the lowest is 3.9%. By contrast, the ABV of wines is typically in the range 11–15%.

## Overview of malting and brewing (Fig. 2.1)

Brewer's yeast *Saccharomyces* can grow on sugar anaerobically by fermenting it to ethanol:

$$C_6H_{12}O_6 \rightarrow 2C_2H_5OH + 2CO_2$$

While *malt* and *yeast* contribute substantially to the character of beers, the quality of beer is at least as much a function of the *water* and, especially, of the *hops* used in its production.

Barley starch supplies most of the sugars from which the alcohol is derived in the majority of the world's beers. Historically, this is because, unlike other cereals, barley retains its husk on threshing and this husk traditionally forms the filter bed through which the liquid extract of sugars is separated in the brewery. Even so, some beers are made largely from wheat while others are from sorghum.

The starch in barley is enclosed in a cell wall and proteins and these wrappings are stripped away in the malting process (essentially a limited germination of the barley grains), leaving the starch largely preserved. Removal of the wall framework softens the grain and makes it more readily milled.

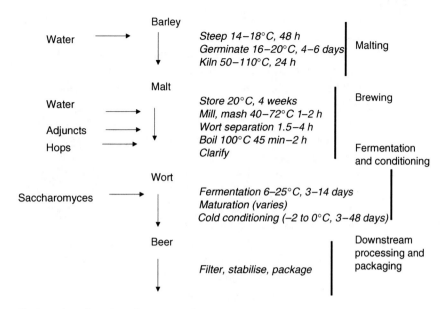

**Fig. 2.1**  Overview of malting and brewing.

Not only that, unpleasant grainy and astringent characters are removed during malting.

In the brewery, the malted grain must first be *milled* to produce relatively fine particles, which are for the most part starch. The particles are then intimately mixed with hot water in a process called *mashing*. The water must possess the right mix of salts. For example, fine ales are produced from waters with high levels of calcium while famous pilsners are from waters with low levels of calcium. Typically mashes have a thickness of three parts water to one part malt and contain a stand at around 65°C, at which temperature the granules of starch are converted by gelatinisation from an indigestible granular state into a 'melted' form that is much more susceptible to enzymatic digestion. The enzymes that break down the starch are called the amylases. They are developed during the malting process, but only start to act once the gelatinisation of the starch has occurred in the mash tun. Some brewers will have added starch from other sources, such as maize (corn) or rice, to supplement that from malt. These other sources are called *adjuncts*. After perhaps an hour of mashing, the liquid portion of the mash, known as *wort*, is recovered, either by straining through the residual *spent grains* or by filtering through plates. The wort is run to the *kettle* (sometimes known as the *copper*, even though they are nowadays fabricated from stainless steel) where it is boiled, usually for around 1 h. *Boiling* serves various functions, including sterilisation of wort, precipitation of proteins (which would otherwise come out of solution in the finished beer and cause cloudiness), and the driving away of unpleasant grainy characters originating in the barley. Many brewers also add some adjunct sugars at this stage, at which most brewers introduce at least a proportion of their *hops*.

The hops have two principal components: resins and essential oils. The resins (so-called $\alpha$-acids) are changed ('isomerised') during boiling to yield iso-$\alpha$-acids, which provide the bitterness to beer. This process is rather inefficient. Nowadays, hops are often extracted with liquefied carbon dioxide and the extract is either added to the kettle or extensively isomerised outside the brewery for addition to the finished beer (thereby avoiding losses due to the tendency of the bitter substances to stick on to yeast). The oils are responsible for the 'hoppy nose' on beer. They are very volatile and if the hops are all added at the start of the boil, then all of the aroma will be blown up the chimney (stack). In traditional lager brewing, a proportion of the hops is held back and only added towards the end of boiling, which allows the oils to remain in the wort. For obvious reasons, this process is called *late hopping*. In traditional ale production, a handful of hops is added to the cask at the end of the process, enabling a complex mixture of oils to give a distinctive character to such products. This is called *dry hopping*. Liquid carbon dioxide can be used to extract oils as well as resins and these extracts can also be added late in the process to make modifications to beer flavour.

After the removal of the precipitate produced during boiling ('hot break', 'trub'), the hopped wort is cooled and *pitched* with yeast. There are many strains of brewing yeast and brewers carefully look after their own strains because of their importance in determining brand identity. Fundamentally brewing yeast can be divided into ale and lager strains, the former type collecting at the surface of the fermenting wort and the latter settling at the bottom of a fermentation (although this differentiation is becoming blurred with modern fermenters). Both types need a little oxygen to trigger off their metabolism, but otherwise the alcoholic *fermentation* is anaerobic. Ale fermentations are usually complete within a few days at temperatures as high as 20°C, whereas lager fermentations at temperatures as low as 6°C can take several weeks. Fermentation is complete when the desired alcohol content has been reached and when an unpleasant butterscotch flavour, which develops during all fermentations, has been mopped up by yeast. The yeast is harvested for use in the next fermentation.

In *traditional ale brewing*, the beer is now mixed with hops, some priming sugars and with isinglass finings from the swim bladders of certain fish, which settle out the solids in the cask.

In *traditional lager brewing*, the 'green beer' is matured by several weeks of cold storage, prior to filtering.

Nowadays, the majority of beers, both ales and lagers, receive a relatively short *conditioning* period after fermentation and before filtration. This conditioning is ideally performed at −1°C or lower (but not so low as to freeze the beer) for a minimum of 3 days, under which conditions more proteins drop out of the solution, making the beer less likely to cloud in the package or glass.

The filtered beer is adjusted to the required carbonation before packaging into cans, kegs, or glass or plastic bottles.

## Barley

Although it is possible to make beer using raw barley and added enzymes (so-called barley brewing), this is extremely unusual. Unmalted barley alone is unsuitable for brewing beer because (1) it is hard and difficult to mill; (2) it lacks most of the enzymes needed to produce fermentable components in wort; (3) it contains complex viscous materials that slow down solid–liquid separation processes in the brewery, which may cause clarity problems in beer and (4) it contains unpleasant raw and grainy characters and is devoid of pleasant flavours associated with malt.

Barley belongs to the grass family. Its Latin name is *Hordeum vulgare*, though this term tends to be retained for six-row barley (discussed later), with *Hordeum distichon* being used for two-row barley. The part of the plant of interest to the brewer is the grain on the ear. Sometimes this is referred to as the seed, but individual grains are generally called kernels or corns. A schematic diagram of a single *barley corn* is shown in Fig. 2.2.

Four components of the kernel are particularly significant:

(1) the embryo, which is the baby plant;
(2) the starchy endosperm, which is the food reserve for the embryo;

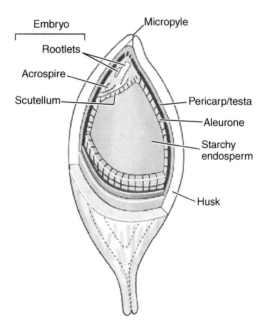

**Fig. 2.2** A barley corn.

(3) the aleurone layer, which generates the enzymes that degrade the starchy endosperm;

(4) the husk (hull), which is the protective layer around the corn. Barley is unusual amongst cereals in retaining a husk after threshing and this tissue is traditionally important for its role as a filter medium in the brewhouse when the wort is separated from spent grains.

The first stage in malting is to expose the grain to water, which enters an undamaged grain solely through the micropyle and progressively hydrates the embryo and the endosperm. This switches on the metabolism of the embryo, which sends hormonal signals to the aleurone layer, triggers that switch on the synthesis of enzymes responsible for digesting the components of the starchy endosperm. The digestion products migrate to the embryo and sustain its growth.

The aim is controlled germination, to soften the grain, remove troublesome materials and expose starch without promoting excessive growth of the embryo that would be wasteful (*malting loss*). The three stages of commercial malting are

(1) *steeping*, which brings the moisture content of the grain to a level sufficient to allow metabolism to be triggered in the grain;

(2) *germination*, during which the contents of the starchy endosperm are substantially degraded ('*modification*') resulting in a softening of the grain;

(3) *kilning*, in which the moisture is reduced to a level low enough to arrest modification.

The embryo and aleurone are both living tissues, but the starchy endosperm is dead. It is a mass of cells, each of which comprises a relatively thin cell wall (approximately 2 μm) inside which are packed many starch granules amidst a matrix of protein (see Fig. 2.3). This starch and protein (and also the cell-wall

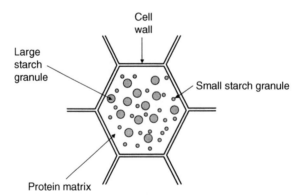

**Fig. 2.3**  A single cell within the starchy endosperm of barley. Only a very small number of the multitude of small and large starch granules are depicted.

materials) are the food reserves for the embryo. However, the brewer's interest in them is as the source of fermentable sugars and assimilable amino acids that the yeast will use during alcoholic fermentation.

The wall around each cell of the starchy endosperm comprises 75% β-glucan, 20% pentosan, 5% protein and some acids, notably acetic acid and the phenolic acid, ferulic acid. The *β-glucan* comprises long linear chains of glucose units joined through β-linkages. Approximately 70% of these linkages are between C-1 and C-4 of adjacent glucosyl units (so-called β 1–4 links, just as in cellulose) and the remainder are between C-1 and C-3 of adjacent glucoses (β 1–3 links, which are *not* found in cellulose) (Fig. 2.4). These 1–3 links disrupt the structure of the β-glucan molecule and make it less ordered, more soluble and digestible than cellulose. Much less is known about the *pentosan* (arabinoxylan, Fig. 2.5) component of the wall, and it is generally believed that it is less easily solubilised and difficult to breakdown when compared with the β-glucan, and that it largely remains in the spent grains after mashing. The cell-wall polysaccharides are problematic because they restrict the yield of extract. They do this either when they are insoluble (by wrapping around the starch components) or when they are solubilised (by restricting the flow of wort from spent grains during wort separation). Dissolved but undegraded β-glucans also increase the viscosity of beer and slow down filtration. They

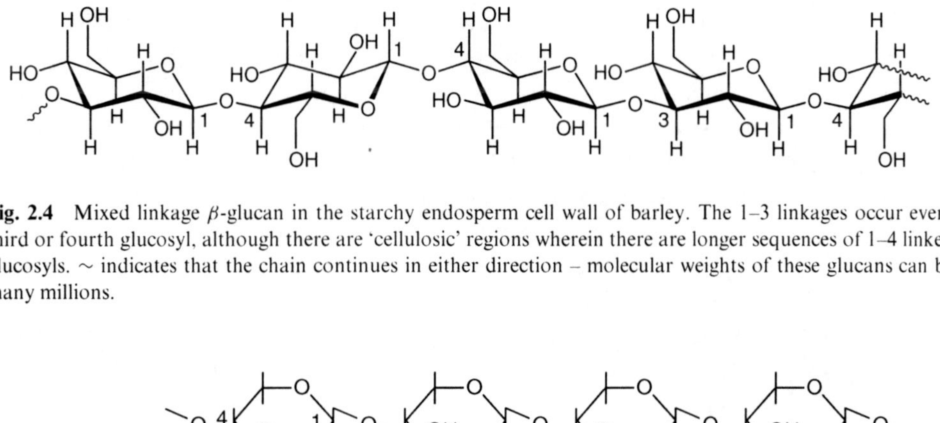

**Fig. 2.4** Mixed linkage β-glucan in the starchy endosperm cell wall of barley. The 1–3 linkages occur every third or fourth glucosyl, although there are 'cellulosic' regions wherein there are longer sequences of 1–4 linked glucosyls. ~ indicates that the chain continues in either direction – molecular weights of these glucans can be many millions.

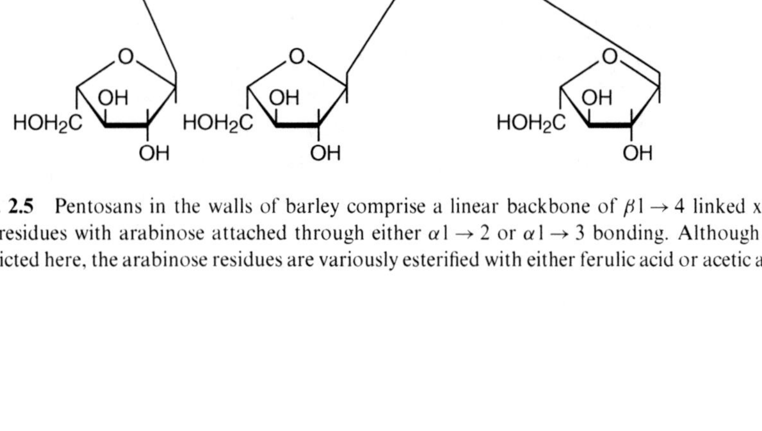

**Fig. 2.5** Pentosans in the walls of barley comprise a linear backbone of β1 → 4 linked xylosyl residues with arabinose attached through either α1 → 2 or α1 → 3 bonding. Although not depicted here, the arabinose residues are variously esterified with either ferulic acid or acetic acid.

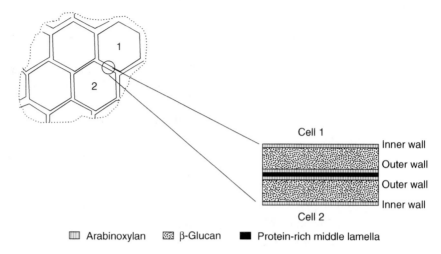

☐☐☐ Arabinoxylan      ▨ β-Glucan      ■ Protein-rich middle lamella

**Fig. 2.6**  Current understanding of the structure of the cell walls of barley endosperm. Walls surrounding adjacent cells are cemented by a protein-rich middle lamella. To this is attached arabinoxylan, within which is the β-glucan.

are prone to drop out of solution as hazes, precipitates or gels. Conversely it has been claimed that β-glucans have positive health attributes for the human, by lowering cholesterol levels and contributing to dietary fibre.

The enzymic breakdown of β-glucan during the germination of barley and later in mashing is in two stages: solubilisation and hydrolysis. Several enzymes (collectively the activity is referred to by the trivial name 'solubilase') may be involved in releasing β-glucan from the cell wall, including esterases that hydrolyse ester bonds believed to cement polysaccharides, perhaps to the protein-rich middle lamella. The most recent evidence, however, is that the pentosan component encloses much of the glucan (Fig. 2.6), and accordingly pentosanases are efficient solubilases. This is despite the observations that pentosans are less digestible than glucans. β-Glucans are hydrolysed by endo-β-glucanases (*endo* enzymes hydrolyse bonds inside a polymeric molecule, releasing smaller units, which are subsequently broken down by *exo* enzymes that chop off one unit at a time, commencing at one end of the molecule). These enzymes convert viscous β-glucan molecules to non-viscous oligosaccharides comprising three or four glucose units. Less well-understood enzymes are responsible for converting these oligosaccharides to glucose. There is little if any β-glucanase in raw barley, it being developed during the germination phase of malting in response to gibberellins. Endo-β-glucanase is extremely sensitive to heat, meaning that it is essential that malt is kilned very carefully to conserve this enzyme if it is necessary that it should complete the task of glucan degradation in the brewhouse. This is especially important if the brewer is using β-glucan-rich adjuncts such as unmalted barley, flaked barley and roasted barley. It is also the reason why brewers often employ a low temperature start to their mashing processes. Alternatively, some brewers add exogenous heat-stable β-glucanases of microbial origin.

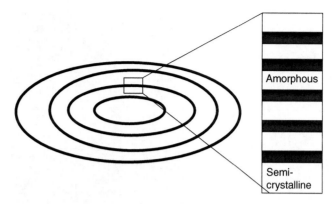

**Fig. 2.7**  The cross-sectional structure of a starch granule.

The *starch* in the cells of the starchy endosperm is in two forms: large granules (approximately 25 μm) and small granules (5 μm). The structure of granules is quite complex, having crystalline and amorphous regions (Fig. 2.7). I address starch later, in the context of mashing.

The *proteins* in the starchy endosperm may be classified according to their solubility characteristics. The two most relevant classes are the *albumins* (water-soluble, some 10–15% of the total) and the *hordeins* (alcohol-soluble, some 85–90% of the total). In the starchy endosperm of barley, the latter are quantitatively the most significant: they are the storage proteins. They need to be substantially degraded in order that the starch can be accessed and amino acids (which will be used by the yeast) generated. Their partial degradation products can also contribute to haze formation via cross-linking with polyphenols. Excessive proteolysis should not occur, however, as some partially degraded protein is required to afford stable foam to beer. Most of the proteolysis occurs during germination rather than subsequent mashing, probably because endogenous molecules that can inhibit the endo-proteinases are kept apart from these enzymes by compartmentalisation in the grain, but when the malt is milled, this disrupts the separation and the inhibitors can now exert their effect. There may be some ongoing protein extraction and precipitation during mashing, and peptides are converted into amino acids at this stage through the action of carboxypeptidases. The endo-peptidases are synthesised during germination in response to gibberellin and they are relatively heat-labile (like the endo-β-glucanases). Substantial carboxypeptidase is present in raw barley and it further increases to abundant levels during germination. It is a heat-resistant enzyme and is unlikely to be limiting. Thus, the extent of protein degradation is largely a function of the extent of proteinase activity during germination.

Much effort is devoted to breeding malting barleys that give high yields of 'extract' (i.e. fermentable material dissolved as wort). The hygiene status of the barley is also very important, and pesticide usage may be important to avoid the risk of infection from organisms such as Fusarium. Barleys may be

Two-row    Six-row

**Fig. 2.8**  Two-row and six-row ears of barley. Photograph courtesy of Dr Paul Schwarz.

two-row, in which only one kernel develops at each node on the ear and it appears as if there is one kernel on either side of the axis of the ear, or six-row in which there are three corns per node (Fig. 2.8). Obviously there is less room for the individual kernels in the latter case and they tend to be somewhat twisted and smaller and therefore less desirable. Farmers are restricted in how much nitrogenous fertiliser they can use because the grain will accumulate protein at the expense of starch in the endosperm, and it is the starch (*ergo* fermentable sugar) that is especially desirable. Maltsters pay a 'malting premium' for the right variety, grown to have the desired level of protein. There must be some protein present, as this is the fraction of the grain which includes the enzymes and which is the origin of amino acids (for yeast metabolism) and foam polypeptide. The amount of protein needed in malt will depend on whether the brewer intends to use some adjunct material as a substitute for malt. For example, corn syrup is a rich source of sugar but not of amino acids, which will need to come from the malt.

Dead grain will not germinate, so batches of barley must pass viability tests.

Most barley in the Northern Hemisphere is sown between January and April and is referred to as Spring Barley. The earlier the sowing, the better the yield and lower the protein levels because starch accumulates throughout the growing season. In locales with mild winters, some varieties (Winter Barleys) are sown in September and October. Best yields of grain are in locales where there is a cool, damp growing season allowing steady growth, and then fine, dry weather at harvest to ripen and dry the grain. Grain grown through very

**Table 2.1**  World production of barley (3-year average, 1998–2000).

| Countries | Production (thousand tons) | Percentage of world production |
|---|---|---|
| World | 132 393 | — |
| Canada | 13 124 | 9.9 |
| Germany | 12 671 | 9.5 |
| Russian Fed. | 11 222 | 8.5 |
| France | 10 036 | 7.5 |
| Spain | 9 871 | 7.4 |
| Turkey | 7 533 | 5.6 |
| USA | 6 908 | 5.2 |
| UK | 6 566 | 5.0 |
| Ukraine | 6 389 | 4.8 |
| Australia | 5 372 | 4.1 |

hot, dry summers is thin, poorly filled and has high nitrogen. Malting barley is grown in many countries (Table 2.1).

Grain arrives at the maltings by road or rail and, as the transport waits, the barley will be weighed and a sample tested for viability, nitrogen content and moisture. Expert evaluation will also provide a view on how clean the sample is in terms of weed content and whether the grain 'smells sweet'. Once accepted, the barley will be cleaned and screened to remove small grain and dust, before passing into a silo, perhaps via a drying operation in areas with damp climates. Grain should be dry to counter infection and outgrowth.

It is essential that the barley store is protected from the elements, yet it must also be ventilated, because barley, like other cereals, is susceptible to various infections, for example, Fusarium, storage fungi such as Penicillium and Aspergillus, Mildew, and pests, for example, aphids and weevils.

Steeping is probably the most critical stage in malting. If homogeneous malt is to be obtained (which will go on to 'behave' predictably in the brewery), then the aim must be to hydrate the kernels in a batch of barley evenly. Steeping regimes are determined on a barley-by-barley basis by small-scale trials but most varieties need to be taken to 42–46%. Apart from water, barley needs oxygen in order to support respiration in the embryo and aleurone. Oxygen access is inhibited if grain is submerged for excessive periods in water, a phenomenon which directly led to the use of interrupted steeping operations. Rather than submerge barley in water and leave it, grain is steeped for a period of time, before removing the water for a so-called 'air-rest' period. Then a further steep is performed and so on. Air rests serve the additional purpose of removing carbon dioxide and ethanol, either of which will suppress respiration. A typical steeping regime may involve an initial steep to 32–38% moisture (lower for more water-sensitive barleys). The start of germination is prompted by an air rest of 10–20 h, followed by a second steep to raise the water content to 40–42%. Emergence of the root tip ('chitting') is encouraged by a second air rest of 10–15 h, before the final steep to the target moisture. The entire steeping operation may take 48–52 h.

Gibberellic acid (GA, itself produced in a commercial fermentation reaction from the fungus *Gibberella*) is added in some parts of the world to supplement the native gibberellins of the grain. Although some users of malt prohibit its use, GA can successfully accelerate the malting process. It is sprayed on to grain at levels between 0.1 and 0.5 ppm as it passes from the last steep to the germination vessel.

The hormones migrate to the aleurone to regulate enzyme synthesis, for the most part to promote the synthesis of enzymes that break down successively β-glucan, protein and starch. The gibberellin first reaches the aleurone nearest to the embryo and therefore, enzyme release is initially into the proximal endosperm. Breakdown of the endosperm ('modification'), therefore, passes in a band from proximal to distal regions of the grain.

Traditionally, steeped barley was spread out to a depth of up to 10 cm on the floors of long, low buildings and germinated for periods up to 10 days. Men would use rakes either to thin out the grain ('the piece') or pile it up depending on whether the batch needed its temperature lowered or raised: the aim was to maintain it at 13–16°C. Very few such floor maltings survive because of their labour intensity, and a diversity of pneumatic (mechanical) germination equipment is now used. Newer germination vessels are circular, made of steel or concrete, with capacities of as much as 500 tons and with turning machinery that is microprocessor-controlled. A modern malting plant is arranged in a tower format, with vessels vertically stacked, steeping tanks uppermost.

Germination in a pneumatic plant is generally at 16–20°C. Once the whole endosperm is readily squeezed out and if the shoot initials (the acrospire) are about three-quarters the length of the grain (the acrospire grows the length of the kernel between the testa and the aleurone and emerges from the husk at the distal end of the corn), then the 'green malt' is ready for kilning.

Through the controlled drying (kilning) of green malt, the maltster is able to

(1) arrest modification and render malt stable for storage;
(2) ensure survival of enzymes for mashing;
(3) introduce desirable flavour and colour characteristics and eliminate undesirable flavours.

Drying should commence at a relatively low temperature to ensure survival of the most heat-sensitive enzymes (enzymes are more resistant to heat when the moisture content is low). This is followed by a progressive increase of temperature to effect the flavour and colour changes (Maillard reaction) and complete drying within the limited turnaround time available (typically under 24 h). There is a great variety of kiln designs, but most modern ones feature deep beds of malt. They have a source of heat for warming incoming air, a fan to drive or pull the air through the bed, together with the necessary loading and stripping systems. The grain is supported on a wedge-wire floor that permits air to pass through the bed, which is likely to be up to 1.2 m deep.

Newer kilns also use 'indirect firing', in that the products of fuel combustion do not pass through the grain bed, but are sent to exhaust, the air being warmed through a heater battery containing water as the conducting medium. Indirect firing arose because of concerns with the role of oxides of nitrogen present in kiln gases that might have promoted the formation of nitrosamines in malt. Nitrosamine levels are now seldom a problem in malt.

Lower temperatures will give malts of lighter colour and will tend to be employed in the production of malts destined for lager-style beers. Higher temperatures, apart from giving darker malts, also lead to a wholly different flavour spectrum. Lager malts give beers that are relatively rich in sulphur compounds, including DMS. Ale malts have more roast, nutty characters. For both lager and ale malts, kilning is sufficient to eliminate the unpleasant raw, grassy and beany characters associated with green malt.

When kilning is complete, the heat is switched off and the grain is allowed to cool before it is stripped from the kiln in a stream of air at ambient temperatures. On its way to steel or concrete hopper-bottomed storage silos, the malt is 'dressed' to remove dried rootlets, which go to animal feed.

Some malts are produced not for their enzyme content but rather for use by the brewer in relatively small quantities as a source of extra colour and distinct types of flavour. These roast malts may also be useful sources of natural antioxidant materials. There is much interest in these products for the opportunities they present for brewing new styles of beer.

## Mashing: the production of sweet wort

Sweet wort is the sugary liquid that is extracted from malt (and other solid adjuncts used at this stage) through the processes of milling, mashing and wort separation. Larger breweries will have raw materials delivered in bulk (rail or road) with increasingly sophisticated unloading and transfer facilities as the size increases. Smaller breweries will have malt, etc. delivered by sack. Railcars may carry up to 80 tons of malt and a truck 20 tons. The conscientious brewer will check the delivery and the vehicle it came in for cleanliness and will representatively sample the bulk. The resultant sample will be inspected visually and smelled before unloading is permitted. Most breweries will spot-check malt deliveries for key analytical parameters to enable them to monitor the quality of a supplier's material against the agreed contractual specification. Grist materials are stored in silos sized according to brewhouse throughput.

### *Milling*

Before malt or other grains can be extracted, they must be milled. Fundamentally the more extensive the milling, the greater the potential there is to extract materials from the grain. However, in most systems for separating wort from spent grains after mashing, the husk is important as a filter medium.

The more intact the husk, the better the filtration. Therefore, milling must be a compromise between thoroughly grinding the endosperm while leaving the husk as intact as possible.

There are fundamentally two types of milling: dry milling and wet milling. In the former, mills may be either roll, disk or hammer. If wort separation is by a lauter tun (discussed later), then a roll mill is used. If a mash filter is installed, then a hammer (or disk) mill may be employed because the husk is much less important for wort separation by a mash filter. Wet milling, which was adopted from the corn starch process, was introduced into some brewing operations as an opportunity to minimise damage to the husk on milling. By making the husk 'soggy', it is rendered less likely to shatter than would a dry husk.

## Mashing

Mashing is the process of mixing milled grist with heated water in order to digest the key components of the malt and generate wort containing all the necessary ingredients for the desired fermentation and aspects of beer quality. Most importantly it is the primary stage for the breakdown of starch.

The starch in the granules is very highly ordered, which tends to make the granules difficult to digest. When granules are heated (in the case of barley starch beyond 55–65°C), the molecular order in the granules is disrupted in a process called gelatinisation. Now that the interactions (even to the point of crystallinity) within the starch have been broken down, the starch molecules become susceptible to enzymic digestion. It is for the purpose of gelatinisation and subsequent enzymic digestion that the mashing process in brewing involves heating.

Although 80–90% of the granules in barley are small, they only account for 10–15% of the total weight of starch. The small granules are substantially degraded during the malting process, whereas degradation of the large granules is restricted to a degree of surface pitting. (This is important, as it is not in the interests of the brewer (or maltster) to have excessive loss of starch, which is needed as the source of sugar for fermentation.)

The starch in barley (as in other plants) is in two molecular forms (Fig. 2.9): amylose, which has very long linear chains of glucose units, and amylopectin, which comprises shorter chains of glucose units that are linked through side chains.

Several enzymes are required for the complete conversion of starch to glucose. $\alpha$-Amylase, which is an endo enzyme, hydrolyses the $\alpha$ 1–4 bonds within amylose and amylopectin. $\beta$-Amylase, an exo enzyme, also hydrolyses $\alpha$ 1–4 bonds, but it approaches the substrate (either intact starch or the lower molecular weight 'dextrins' produced by $\alpha$-amylase) from the non-reducing end, chopping off units of two glucoses (i.e. molecules of maltose). Limit dextrinase is the third key activity, attacking the $\alpha$ 1–6 side chains in amylopectin.

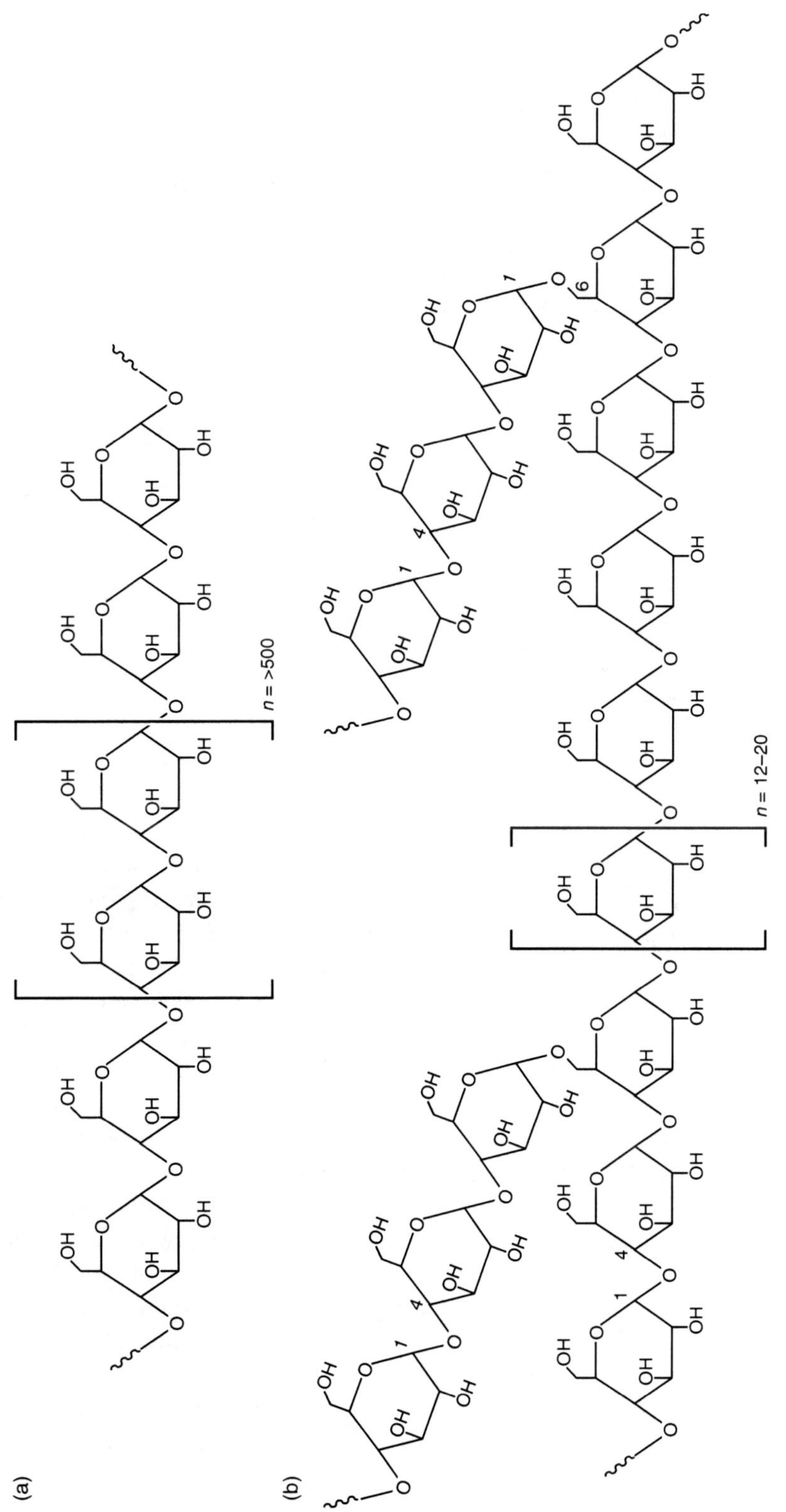

**Fig. 2.9** (Continued).

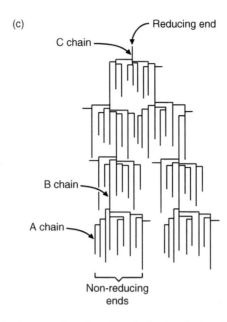

**Fig. 2.9** (a) The basic structure of amylose. Not depicted is the fact that it assumes a helical structure. (b) The basic structure of amylopectin. The individual linear chains adopt a helical conformation. (c) The different types of chain in amylopectin. The different layers in the starch granule result from the ordering of these molecules, interacting with amylose.

$\alpha$-Amylase develops during the germination phase of malting. It is extremely heat resistant, and also present in very high activity; therefore, it is capable of extensive attack, not only on the starch from malt but also on that from adjuncts added in quantities of 50% or more. $\beta$-Amylase is already present in the starchy endosperm of raw barley, in an inactive form through its association with protein Z. It is released during germination by the action of a protease (and perhaps a reducing agent). $\beta$-Amylase is considerably more heat-labile than $\alpha$-amylase, and will be largely destroyed after 30–45 min of mashing at 65°C. Limit dextrinase is similarly heat sensitive. Furthermore, it is developed much later than the other two enzymes, and germination must be prolonged if high levels of this enzyme are to be developed. It is present in several forms (free and bound): the bound form is both synthesised and released during germination. Like the proteinases, there are endogenous inhibitors of limit dextrinase in grain, and this is probably the main factor which determines that some 20% of the starch in most brews is left in the wort as non-fermentable dextrins. Although it is possible to contrive operations that will allow greater conversion of starch to fermentable sugar, in practice, many brewers seeking a fully fermentable wort add a heat-resistant glucoamylase (e.g. from *Aspergillus*) to the mash (or fermenter). This enzyme has an exo action like $\beta$-amylase, but it chops off individual glucose units.

There are several types of mashing which can broadly be classified as *infusion mashing, decoction mashing* and *temperature-programmed mashing*.

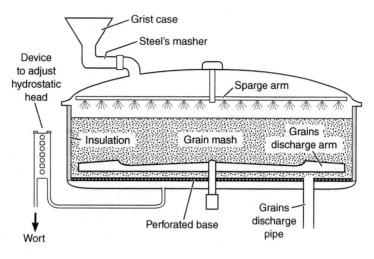

**Fig. 2.10** A mash tun.

Whichever type of mashing is employed, the vessels these days are almost exclusively fabricated from stainless steel (once they were copper). What stainless steel loses in heat transfer properties is made up for in its toughness and ability to be cleaned thoroughly by caustic and acidic detergents.

Irrespective of the mashing system, most mashing systems (apart from wet milling operations) incorporate a device for mixing the milled grist with water (which some brewers call 'liquor'). This device, the 'pre-masher', can be of various designs, the classic one being the Steel's masher, which was developed for the traditional infusion mash tun (Fig. 2.10).

*Infusion mashing* is relatively uncommon, but still championed by traditional brewers of ales. It was designed in England to deal with well-modified ale malts that did not require a low temperature start to mashing in order to deal with residual cell-wall material ($\beta$-glucans). Grist is mixed with water (a typical ratio would be three parts solid to one part water) in a Steel's masher en route to the preheated mash tun, with a single holding temperature, typically 65°C, being employed. This temperature facilitates gelatinisation of starch and subsequent amylolytic action. At the completion of this 'conversion', wort is separated from the spent grains in the same vessel, which incorporates a false bottom and facility to regulate the hydrostatic pressure across the grains bed. The grist is sparged to enable leaching of as much extract as possible from the bed.

*Decoction mashing* was designed on the mainland continent of Europe to deal with lager malts which were less well-modified than ale malts. Essentially it provides the facility to start mashing at a relatively low temperature, thereby allowing hydrolysis of the $\beta$-glucans present in the malt, followed by raising the temperature to a level sufficient to allow gelatinisation of starch and its subsequent enzymic hydrolysis. The manner by which the temperature increase was achieved was by transferring a portion of the initial mash to a

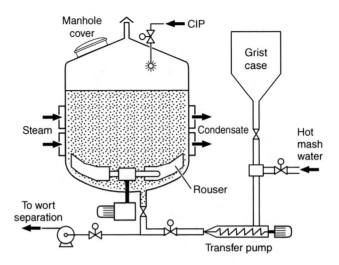

**Fig. 2.11** A mash converter.

separate vessel where it was taken to boiling and then returned to the main mash, leading to an increase in temperature. This is a rather simplified version of the process, which traditionally involved several steps of progressive temperature increase.

*Temperature-programmed mashing.* Although there are some adherents to the decoction-mashing protocol, most brewers nowadays employ the related but simpler temperature-programmed mashing. Again, the mashing is commenced at a relatively low temperature, but subsequent increases in temperature are effected in a single vessel (Fig. 2.11) by employing steam-heated jackets around the vessel to raise the temperature of the contents, which are thoroughly mixed to ensure even heat transfer. Mashing may commence at 45–50°C, followed by a temperature rise of 1°C.min$^{-1}$ until the conversion temperature (63–68°C) is reached. The mash will be held for perhaps 50 min to 1 h, before raising the temperature again to the sparging temperature (76–78°C). High temperatures are employed at the end of the process to arrest enzymic activity, to facilitate solubilisation of materials and to reduce viscosity, thereby allowing more rapid liquid–solid separation.

## Adjuncts

The decision whether to use an adjunct or not is made on the basis of cost (does it represent a cost advantageous source of extract, compared to malted barley?) and quality (does the adjunct provide a quality benefit, in respect of flavour, foam or colour?). Liquid adjuncts (sugars /syrups) are added in the wort boiling stage (discussed later). A series of solid adjuncts may be added at the mashing stage because they depend on the enzymes from malt to digest their component macromolecules. Solid adjuncts may be based on other cereals as well as barley: wheat, corn (maize), rice, oats, rye and sorghum.

**Table 2.2**  Gelatinisation temperatures of starches from different cereals.

| Source | Gelatinisation temperature (°C) |
|---|---|
| Barley | 61–62 |
| Corn | 70–80 |
| Oats | 55–60 |
| Rice | 70–80 |
| Rye | 60–65 |
| Sorghum | 70–80 |
| Wheat | 52–54 |

In turn, these adjuncts can be in different forms: raw cereal (barley, wheat); raw grits (corn, rice, sorghum); flaked (corn, rice, barley, oats); micronised or torrefied (corn, barley, wheat); flour/starch (corn, wheat, sorghum) and malted (apart from barley this includes wheat, oats, rye, and sorghum).

A key aspect of solid adjuncts is the gelatinisation temperature of the starch (Table 2.2). A higher gelatinisation temperature for corn, rice and sorghum means that these cereals need treatment at higher temperatures than do barley, oats, rye or wheat. If the cereal is in the form of grits (produced by the dry milling of cereal in order to remove outer layers and the oil-rich germ), then it needs to be 'cooked' in the brewhouse. Alternatively, the cereal can be pre-processed by intense heat treatment in a micronisation or flaking operation. In the former process, the whole grain is passed by conveyor under an intense heat source (260°C), resulting in a 'popping' of the kernels (cf. puffed breakfast cereals). In flaking, grits are gelatinised by steam and then rolled between steam-heated rollers. Flakes are not required to be milled in the brewhouse, but micronised cereals are.

Cereal cookers employed for dealing with grits are made of stainless steel and incorporate an agitator and steam jackets. The adjunct is delivered from a hopper and the adjunct will be mixed with water at a rate of perhaps 15 kg per hL of water. The adjunct will be mixed with 10–20% of malt as a source of enzymes. The precise temperature employed in the cooker will depend on the adjunct and the preferences of the brewer. Following cooking, the adjunct mash is likely to be taken to boiling and then mixed with the main mash (at its mashing-in temperature), with the resultant effect being the temperature rise to conversion for the malt starch (cf. decoction mashing). This is sometimes called 'Double mashing'.

## Wort separation

Traditionally, recovering wort from the residual grains in the brewery is perhaps the most skilled part of brewing. Not only is the aim to produce a wort with as much extract as possible, but many brewers prefer to do this such that the wort is 'bright', that is, not containing many insoluble particles which may present difficulties later. All this needs to take place within a time window, for the mashing vessel must be emptied in readiness for the next brew.

Irrespective of the system employed for mash separation (traditional infusion mash tun, lauter tun, or mash filter), the science dictating rate of liquid recovery is the same and is defined by Darcy's equation:

$$\text{Rate of liquid flow} = \frac{\text{Pressure} \times \text{bed permeability} \times \text{filtration area}}{\text{Bed depth} \times \text{wort viscosity}}$$

And so the wort will be recovered more quickly if the device used to separate the wort has a large surface area, is shallow and if a high pressure can be employed to force the liquid through. The liquid should be of as low viscosity as possible, as less viscous liquids flow more readily. Also the bed of solids should be as permeable as possible. Perhaps the best analogy here is to sand and clay. Sand comprises relatively large particles around which a liquid will flow readily. To pass through the much smaller particles of clay, though, water has to take a much more circuitous route and it is held up. The particle sizes in a bed of grains depend on certain factors, such as the fineness of the original milling and the extent to which the husk survived milling (discussed earlier). Furthermore, a layer (teig or oberteig) collects on the surface of a mash, this being a complex of certain macromolecules, including oxidatively cross-linked proteins, lipids and cell-wall polysaccharides, and this layer has a very fine size distribution analogous to clay. (The oxidative cross-linking of the proteins is exactly akin to that involved in bread dough – see Fig. 12.3). However, particle size also depends on the temperature, and it is known that at the higher temperatures used for wort separation (e.g. 78°C), there is an agglomeration of very fine particles into larger ones past which wort will flow more quickly.

### Lauter tun

Generally this is a straight-sided round vessel with a slotted or wedged wire base and run-off pipes through which the wort is recovered (Fig. 2.12). Within the vessel there are arms that can be rotated about a central axis. These arms carry vertical knives that are used as appropriate to slice through the grains bed and facilitate run-off of the wort. Water can be sparged onto the grain to ensure collection of all the desired soluble material. The spent grains are shipped off site to be used as cattle food.

### Mash filters

Increasingly, modern breweries use mash filters. These operate by using plates of polypropylene to filter the liquid wort from the residual grains (Fig. 2.13). Accordingly, the grains serve no purpose as a filter medium and their particle sizes are irrelevant. The high pressures that can be used in the squeezing of the plates together overcome the reduced permeability due to smaller particle

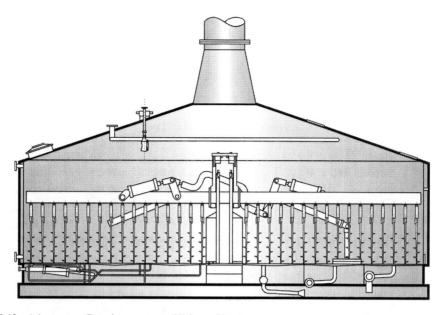

**Fig. 2.12**   A lauter tun. Drawing courtesy of Briggs of Burton.

**Fig. 2.13**   A mash filter. Photograph courtesy of Briggs of Burton.

sizes (the sand versus clay analogy used earlier). Furthermore, the grains bed depth is particularly shallow (2–3 in.), being nothing more than the distance between the adjacent plates.

# Water

Since water represents at least 90% of the composition of most beers, it will clearly have a major direct impact on the product, particularly in terms of flavour and clarity. The nature of the water, however, exerts its influence much earlier in the process, through the impact of the salts it contains on enzymic and chemical processes, through the determination of pH, etc.

Water in breweries comes either from wells owned by the brewer (cf. the famous water of Burton-on-Trent in England or Pilsen in the Czech Republic) or from municipal supplies; especially in the latter instance, the water will be subjected to clean-up procedures, such as charcoal filtration, to eliminate undesirable taints and colours.

The ionic composition of the water in four brewing centres is given in Table 2.3. The water in Burton is clearly very hard, both permanent and temporary. By contrast, the water in Pilsen is extremely soft. It is clear that the nature of the water has had some impact on the quality of the different beer styles traditionally produced in these two centres; however, the rationale for the differences is less than fully satisfactorily explained.

The water composition can be adjusted, either by adding or by removing ions. Thus, calcium levels may be increased in order to promote the precipitation of oxalic acid as oxalate, to lower the pH by reaction with phosphate ions ($3Ca^{2+} + 2HPO_4^{2-} \rightarrow Ca_3(PO_4)_2 + 2H^+$) and to promote amylase action. (The optimum pH for mashing is between 5.2 and 5.4.) The alkalinity of water used for sparging (alkalinity is largely determined by the content of carbonate and bicarbonate) may be reduced to less than 50 ppm in order to limit the extraction of tannins. Ions such as iron and copper must be as low as possible to preclude oxidation. Furthermore, water may need to be of different standards for different purposes. The microbiological status of water used for slurrying yeast or for use downstream generally is important. Water used for diluting high-gravity streams must be of low oxygen content, and its ionic composition will be critical. When ions need to be removed, the likeliest approach is ion-exchange resin technology.

**Table 2.3**  Ionic composition (mg $L^{-1}$) of water.

| Component | Burton | Pilsen | Dublin | Munich |
|---|---|---|---|---|
| Calcium | 352 | 7 | 119 | 80 |
| Magnesium | 24 | 8 | 4 | 19 |
| Sulphate | 820 | 6 | 54 | 6 |
| Chloride | 16 | 5 | 19 | 1 |
| Bicarbonate | 320 | 37 | 319 | 333 |

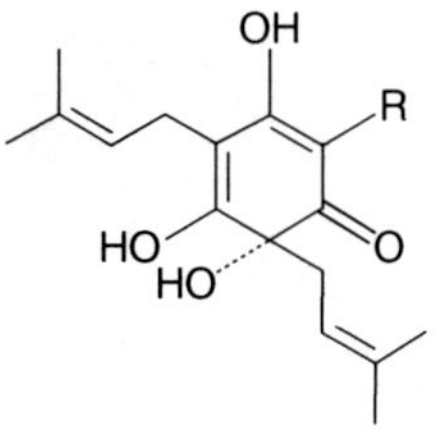

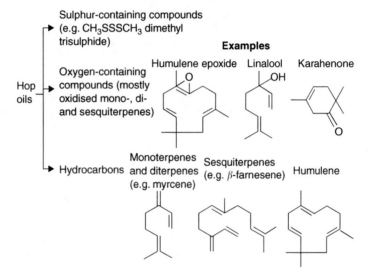

| Name | Side chain (R) |
| --- | --- |
| Humulone | —CO·CH$_2$·CH(CH$_3$)$_2$ isovaleryl |
| Cohumulone | —CO·CH(CH$_3$)$_2$ isobutyryl |
| Adhumulone | —CO·CH(CH$_3$)·CH$_2$·CH$_3$ 2–methylbutyryl |

**Fig. 2.14**  Hop resins.

**Fig. 2.15**  Hop oils.

# Hops

The hop, *Humulus lupulus*, is rich in resins (Fig. 2.14) and oils (Fig. 2.15), the former being the source of bitterness, the latter the source of aroma. The hop is remarkable amongst agricultural crops in that essentially its sole outlet is for brewing. Hops are grown in all temperate regions of the world, with approximately one-third coming from Germany.

Hops are hardy, climbing herbaceous perennial plants grown in gardens using characteristic string frameworks to support them. It is only the female plant that is cultivated, as it is the one that develops the hop cone (Fig. 2.16). Their rootstock remains in the ground year on year and is spaced in an appropriate fashion for effective horticultural procedures (e.g. spraying by tractors passing between rows). In recent years, so-called dwarf varieties have been

**Fig. 2.16**   Hop cones. Photograph courtesy of Yakima Chief.

bred, which retain the bittering and aroma potential of 'traditional' hops but which grow to a shorter height (6–8 ft as opposed to twice as big). As a result, they are much easier to harvest and there is less wastage of pesticide during spraying. Dwarf hop gardens are also much cheaper to establish.

Hops are susceptible to a wide range of diseases and pests. The most serious problems come from *Verticillium* wilt, downy mildew, mould and the damson-hop aphid. Varieties differ in their susceptibility to infestation and have been progressively selected on this basis. Nonetheless, it is frequently necessary to apply pesticides, which are always stringently evaluated for their influence on hop quality, for any effect they may have on the brewing process and, of course, for their safety.

Hops are generally classified into two categories: aroma hops and bittering hops. All hops are capable of providing both bitterness and aroma. Some hops, however, such as the Czech variety Saaz, have a relatively high ratio of oil to resin and the character of the oil component is particularly prized. Such varieties command higher prices and are known as *aroma varieties*. They are seldom used as the sole source of bitterness and aroma in a beer: a cheaper, higher $\alpha$-acid hop (a *bittering variety*) is used to provide the bulk of the bitterness, with the prized aroma variety added late in the boil for the contribution of its own unique blend of oils. Those brewers requiring hops solely as a source of bitterness may well opt for a cheaper variety, ensuring its use early in the kettle boil so that the provision of bitterness is maximised and unwanted aroma is driven off.

The use of whole cone hops is comparatively uncommon nowadays. Many brewers use hops that have been hammer-milled and then compressed into pellets. In this form they are more stable, more efficiently utilised and do not present the brewer with the problem of separating out the vegetative parts of the hop plant. Some use hop extracts that are derived by dissolving the resins in liquid carbon dioxide, followed by a chemical isomerisation if the bitterness is to be added to the finished beer rather than in the boiling stage. Recent years have been marked by an enormous increase in the use of such pre-isomerised extracts after they have been modified by reduction. One of the side chains on the iso-$\alpha$-acids is susceptible to cleavage by light; it then reacts with traces of sulphidic materials in beer to produce 2-methyl-3-butene-1-thiol (MBT), a substance that imparts an intensely unpleasant skunky character to beer. If the side chain is reduced, it no longer produces MBT. For this reason, beers that are destined for packaging in green or clear glass bottles are often produced using these modified bitterness preparations, which have the added advantage of possessing increased foam-stabilising and antimicrobial properties.

## Wort boiling and clarification

The boiling of wort serves various functions, primary amongst which are the isomerisation of the hop resins ($\alpha$-acids) to the more soluble and bitter iso-$\alpha$-acids, sterilisation, the driving off of unwanted volatile materials, the precipitation of protein/polyphenol complexes (as 'hot break' or 'trub') and concentration of the wort. The extent of wort boiling is normally described in terms of percentage evaporation. Water is usually boiled off at a rate of about $4\% \, h^{-1}$ and the duration of boiling is likely to be 1–2 h. Brew kettles are sometimes referred to as 'coppers', reflecting the original metal from which they were fabricated (Fig. 2.17). These days they are usually made from stainless steel. Certain fining materials (e.g. a charged polysaccharide from Irish Moss) may be added to promote protein precipitation. This is the stage at which liquid sugar adjunct can be added (Table 2.4). Sugars added in the kettle are called 'wort extenders': they present the opportunity to increase the extract from a brewhouse without investment in extra mashing vessels and wort separation devices. Most sugars are derived from either corn or sugar cane. In the latter case, the principal sugar is either sucrose or fructose plus glucose if the product has been 'inverted'. There are many different corn sugar products, differing in their degree of hydrolysis and therefore fermentability. Through the controlled use of acid but increasingly of starch-degrading enzymes, the supplier can produce preparations with a full range of fermentabilities depending on the needs of the brewer: from 100% glucose through to high dextrin.

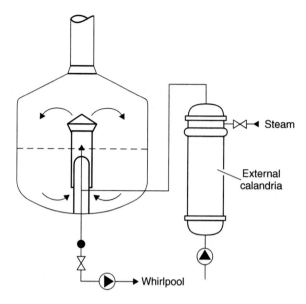

**Fig. 2.17**   Kettle. Wort in this design is siphoned through the external heating device (calandria), thus ensuring an efficient and highly turbulent boil.

**Table 2.4**   Brewing sugars.

| Type | Carbohydrate distribution (%) |
| --- | --- |
| Cane | Sucrose predominantly |
| Invert | Glucose (50), fructose (50) |
| Dextrose | Glucose (100) |
| High conversion (acid + enzyme) | Glucose (88), maltose (4), maltotriose (2), dextrin (6) |
| Glucose chips | Glucose (84), maltose (1), maltotriose (2), dextrin (13) |
| Maize syrup | Glucose (45), maltose (38), maltotriose (3), dextrin (14) |
| Very-high maltose | Glucose (5), maltose (70), maltotriose (10), dextrin (15) |
| High conversion (acid) | Glucose (31), maltose (18), maltotriose (13), dextrin (38) |
| High maltose | Glucose (10), maltose (60), dextrin (30) |
| Low conversion | Glucose (12), maltose (10), maltotriose (10), dextrin (68) |
| Maltodextrin | Maltose (1.5), maltotriose (1.5), dextrin (95) |
| Malt extract | Comparable to brewer's wort – also contains nitrogenous components |

The products dextrose through maltodextrin are customarily derived by the selective hydrolysis of corn (maize)-derived starch by acid and enzymes to varying extents. Derived from *Pauls Malt Brewing Room Book* (1998–2000). Bury St Edmunds: Moreton Hall Press.

After boiling, wort is transferred to a clarification device. The system employed for removing insoluble material after boiling depends on the way in which the hopping was carried out. If whole hop cones are used, clarification is through a hop jack (hop back), which is analogous to a lauter tun, but in this case the bed of residual hops constitutes the filter medium. If hop pellets or extracts are used, then the device of choice is the whirlpool, a cylindrical vessel, into which hot wort is transferred tangentially through an opening

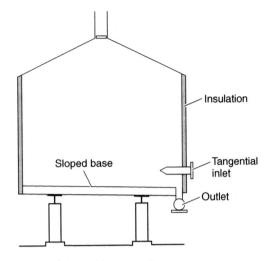

**Fig. 2.18**  A 'whirlpool' (hot wort residence vessel).

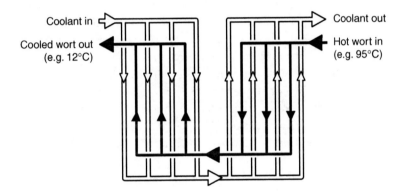

**Fig. 2.19**  A heat exchanger.

0.5–1 m above the base (Fig. 2.18). The wort is set into a rotational flux, which forces trub to a pile in the middle of the vessel.

## Wort cooling

Almost all cooling systems these days are of the stainless steel plate heat exchanger type, sometimes called 'paraflows' (Fig. 2.19). Heat is transferred from the wort to a coolant, either water or glycol depending on how low the temperature needs to be taken. At this stage, it is likely that more material will precipitate from solution ('cold break'). Brewers are divided on whether they feel this to be good or bad for fermentation and beer quality. The presence of this break certainly accelerates fermentation and, therefore, it will directly influence yeast metabolism. As in so much of brewing, the aim should be consistency: either consistently 'bright worts' or ones containing a relatively consistent level of trub.

## Yeast

Brewing yeast is *Saccharomyces cerevisiae* (ale yeast) or *Saccharomyces pastorianus* (lager yeast). There are many separate strains of brewing yeast, each of which is distinguishable phenotypically [e.g. in the extent to which it will ferment different sugars, or in the amount of oxygen it needs to prompt its growth, or in the amounts of its metabolic products (i.e. flavour spectrum of the resultant beer), or its behaviour in suspension (top versus bottom fermenting, flocculent or non-flocculent)] and genotypically, in terms of its DNA fingerprint.

The fundamental differentiation between ale and lager strains is based on the ability or otherwise to ferment the sugar melibiose (Fig. 2.20): ale strains cannot whereas lager strains can because they produce the enzyme (α-galactosidase) necessary to convert melibiose into glucose and galactose. Ale yeasts also move to the top of open fermentation vessels and are called top-fermenting yeasts. Lager yeasts drop to the bottom of fermenters and are termed bottom-fermenting yeasts. Nowadays it is frequently difficult to make this differentiation, when beers are widely fermented in similar types of vessel (deep cylindro-conical tanks) which tend to equalise the way in which yeast behave in suspension.

We considered yeast structure in Chapter 1. When presented with wort, yeast encounters a selection of carbohydrates which, for a typical all-malt wort, will approximate to maltose (45%), maltotriose (15%), glucose (10%), sucrose (5%), fructose (2%) and dextrin (23%). The dextins (maltotetraose and larger) are unfermentable. The other sugars will ordinarily be utilised in the sequence sucrose, glucose, fructose, maltose, and lastly maltotriose, though there may be some overlap (Fig. 2.21). Sucrose is hydrolysed by an enzyme (invertase) released by the yeast outside the cell, and then the glucose and fructose enter the cell to be metabolised. Maltose and maltotriose also

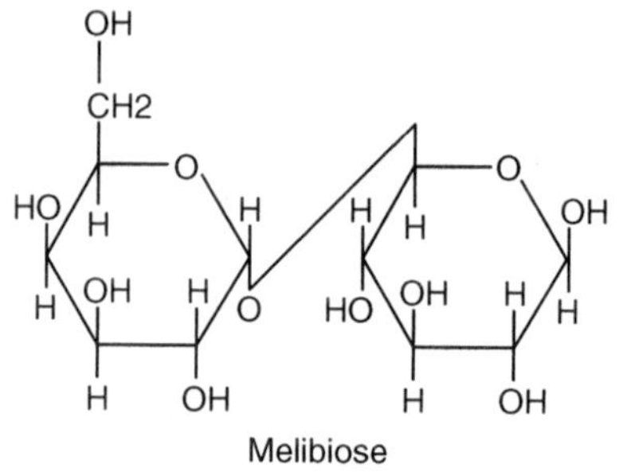

Fig. 2.20   Melibiose.

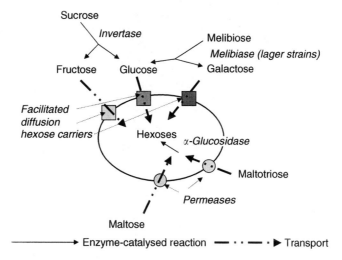

**Fig. 2.21** The uptake of sugars by brewing yeast.

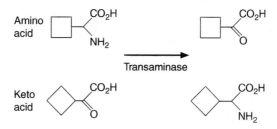

**Fig. 2.22** The principle of transamination.

enter, through the agency of specific permeases. Inside the cell they are broken down into glucose by an $\alpha$-glucosidase. Glucose represses the maltose and maltotriose permeases.

The principal route of sugar utilisation in the cell is the EMP pathway of glycolysis (see Chapter 1). Brewing yeast derives most of the nitrogen it needs for synthesis of proteins and nucleic acids from the amino acids in the wort. A series of permeases is responsible for the sequential uptake of the amino acids. It is understood that the amino acids are transaminated to keto acids and held within the yeast until they are required, when they are transaminated back into the corresponding amino acid (Figs 2.22 and 2.23). The amino acid spectrum and level in wort (free amino nitrogen, FAN) is significant as it influences yeast metabolism leading to flavour-active products.

Oxygen is needed by the yeast to synthesise the unsaturated fatty acids and sterols it needs for its membranes. This oxygen is introduced at the wort cooling stage in the quantities that the yeast requires – but no more, because excessive aeration or oxygenation promotes excessive yeast growth, and the more yeast is produced in a fermentation, the less alcohol will be produced. Different yeasts need different amounts of oxygen.

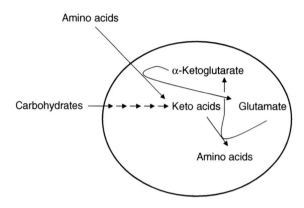

**Fig. 2.23**    Transamination as part of the metabolism of amino acids by yeast.

Yeast uses its stored reserves of carbohydrate in order to fuel the early stages of metabolism when it is pitched into wort, for example, the synthesis of sterols. There are two principal reserves: glycogen and trehalose. Glycogen is similar in structure to the amylopectin fraction of barley starch. Trehalose is a disaccharide comprising two glucoses linked with an $\alpha$-1,1 bond between their reducing carbons. The glycogen reserves of yeast build up during fermentation and it is important that they are conserved in the yeast during storage between fermentations. Trehalose may feature as more of a protection against the stress of starvation. It certainly seems to help the survival of yeast under dehydration conditions employed for the storage and shipping of dried yeast.

Pure yeast culture was pioneered by Hansen at Carlsberg in 1883. By a process of dilution, he was able to isolate individual strains and open up the possibility of selecting and growing separate strains for specific purposes. Nowadays brewers maintain their own pure yeast strains. While it is still a fact that some brewers simply use the yeast grown in one fermentation to 'pitch' the following fermentation, and that they have done this for many tens of years, it is much more usual for yeast to be repropagated from a pure culture every 4–6 generations. (When brewers talk of 'generations', they mean successive fermentations; strictly speaking, yeast advances a generation every time it buds, and therefore there are several generations during any individual fermentation.)

Large quantities of yeast are needed to pitch commercial-scale fermentations. They need to be generated by successive scale-up growth from the master culture (Fig. 2.24). Higher yields are possible if fed-batch culture is used. This is the type of procedure used in the production of baker's yeast. It takes advantage of the Crabtree effect, in which high concentrations of sugar drive the yeast to use it fermentatively rather than by respiration. When yeast grows by respiration, it captures much more energy from the sugar and therefore produces much more cell material. In fed-batch culture, the amount of sugar made available to the yeast at any stage is low. Together with the high levels of oxygen in a well-aerated system, the yeast respires and grows substantially.

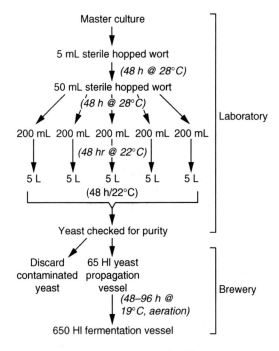

Master culture

5 mL sterile hopped wort

*(48 h @ 28°C)*

50 mL sterile hopped wort

*(48 h @ 28°C)*

200 mL  200 mL  200 mL  200 mL  200 mL

*(48 hr @ 22°C)*

5 L      5 L      5 L      5 L      5 L

(48 h/22°C)

Laboratory

Yeast checked for purity

Discard          65 Hl yeast
contaminated     propagation
yeast            vessel

*(48–96 h @ 19°C, aeration)*

650 Hl fermentation vessel

Brewery

**Fig. 2.24** Yeast propagation. After MacDonald *et al.*. (1984).

The sugar is 'dribbled in' and the end result is a far higher yield of biomass, perhaps four-fold more than is produced when the sugar is provided in a single batch at the start of fermentation.

The majority of brewing yeasts are resistant to acid (pH 2.0–2.2) and so the addition of phosphoric acid to attain this pH is very effective in killing bacteria with which yeast may become progressively contaminated from fermentation to fermentation. Many brewers use such an acid washing of yeast between fermentations.

There are two key indices of yeast health: viability and vitality. Both should be high if a successful fermentation is to be achieved. Viability is a measure of whether a yeast culture is alive or dead. While microscopic inspection of a yeast sample is useful as a gross indicator of that culture (e.g. presence of substantial infection), quantitative evaluation of viability needs a staining test. The most common is the use of methylene blue: viable yeast decolourises it, dead cells do not. Although a yeast cell may be living, it does not necessarily mean that it is healthy. Vitality is a measure of how healthy a yeast cell is. Many techniques have been advanced as an index of vitality, but none has been accepted as definitive.

Preferably yeast is stored in a readily sanitised room that can be cleaned efficiently and which is supplied with a filtered air supply and possesses a pressure higher than the surroundings in order to impose an outwards vector of contaminants. Ideally it should be at or around 0°C. Even if storage is not in such a room, the tanks must be rigorously cleaned, chilled to 0–4°C and have the facility for gently rousing (mixing) to avoid hot spots. Yeast is stored in slurries ('barms') of 5–15% solids under 6 in. of beer, water or potassium phosphate solution. An alternative procedure is to press the yeast and store it at 4°C in a cake form (20–30% dry solids). Pressed yeast may be held for about 10 days, water slurried and beer slurried for 3–4 weeks and slurries in 2% phosphate, pH 5 for 5 weeks.

Brewers seeking to ship yeast normally transport cultures for re-propagation at the destination. However, greater consistency is achieved when it is feasible to propagate centrally and ship yeast for direct pitching. Such yeast must be contaminant-free and of high viability and vitality, washed free from fermentable material and cold (0°C). The longer the distance, the greater the recommendation for low moisture pressed cake.

Apart from the importance of pitching yeast of good condition, it is also important that the amount pitched is in the correct quantity. The higher the pitching rate, the more rapid the fermentation. As the pitching rate increases, initially so too does the amount of new biomass synthesised, until at a certain rate, the amount of new yeast synthesised declines. The rate of attenuation and the amount of growth directly impacts the metabolism of yeast and the levels of its metabolic products (i.e. beer flavour), hence the need for control. Yeast can be quantified by weight or cell number. Typically some $10^7$ cells per mL will be pitched for wort of 12°Plato (1.5–2.5 g pressed weight per L). At such a pitching rate, lager yeast will divide 4–5 times in fermentation. Yeast numbers can be measured using a haemocytometer, which is a counting chamber loaded onto a microscope slide. It is possible to weigh yeast or to centrifuge it down in pots which are calibrated to relate volume to mass, but in these cases it must be remembered that there are usually other solid materials present, for example, trub.

Another procedure that has come into vogue is the use of capacitance probes that can be inserted in-line. An intact and living yeast cell acts as a capacitor and gives a signal whereas dead ones (or insoluble materials) do not. The device is calibrated against a cell number (or weight) technique and therefore allows the direct read-out of the amount of viable yeast in a slurry. Other in-line devices quantify yeast on the basis of light scatter.

## Brewery fermentations

Primary fermentation is the fermentation stage proper in which yeast, through controlled growth, is allowed to ferment wort to generate alcohol and the desired spectrum of flavours. Increasingly brewery fermentations are conducted in cylindro-conical vessels (Fig. 2.25). The fermentation is regulated by control of several parameters, notably the starting strength of the wort (°Plato, which approximates to percentage sugar by weight, or Brix), the amount of viable yeast ('pitching rate'), the quantity of oxygen introduced and the temperature. Fermentation is monitored by measuring the

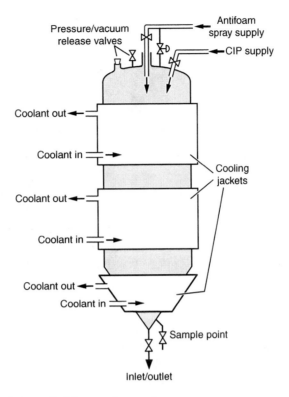

Coolant out

Coolant in

Coolant out

Coolant in

Coolant out

Coolant in

Pressure/vacuum
release valves

Antifoam
spray supply

CIP supply

Cooling
jackets

Sample point

Inlet/outlet

**Fig. 2.25**   A cylindro-conical fermentation vessel.

decrease in specific gravity (alcohol has a much lower specific gravity than sugar).

Ales are generally fermented at a higher temperature (15–20°C) than lagers (6–13°C) and therefore attenuation (the achievement of the finished specific gravity) is achieved more rapidly. Thus, an ale fermenting at 20°C may achieve attenuation gravity in 2 days, whereas a lager fermented at 8.5°C may take 10 days. The temperature has a substantial effect on the metabolism of yeast, and the levels of a flavour substance like iso-butanol will be 16.5 and $7 \, \text{mg L}^{-1}$, respectively, for the ale and the lager. Some brewers add zinc (e.g. 0.2 ppm) to promote yeast action–it is a cofactor for the enzyme alcohol dehydrogenase. During fermentation, the pH falls because yeast secretes organic acids and protons. A diagram depicting the time course of fermentation can be found in Fig. 2.26.

Surplus yeast will be removed at the end of fermentation, either by a process such as 'skimming' for a traditional square fermenter employing top fermenting yeast, or from the base of a cone in a cylindro-conical vessel. This is not only to preserve the viability and vitality of the yeast, but also to circumvent the autolysis and secretory tendencies of yeast that will be to the detriment of flavour and foam. There will still be sufficient yeast in the beer to effect the secondary fermentation.

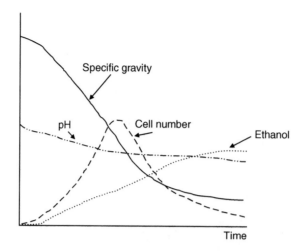

**Fig. 2.26**    Changes occurring during a brewery fermentation.

The 'green' beer produced by primary fermentation needs to be 'conditioned', in respect of establishment of a desired carbon dioxide content and refinement of the flavour. This is called secondary fermentation. Above all at this stage, there needs to be the removal of an undesirable butter-scotch flavour character due to substances called vicinal diketones (VDKs; discussed later). Traditionally it is the lager beers fermented at lower temperatures that have needed the more prolonged maturation (storage: 'lagering') in order to refine their flavour and develop carbonation. The latter depends on the presence of sugars, either those (perhaps 10%) which the brewer ensures are residual from the primary fermentation or those introduced in the 'krausening' process, in which a proportion of freshly fermenting wort is added to the maturing beer. Many brewers are unconvinced by the need for prolonged storage periods (other than for its strong marketing appeal) and they tend to combine the primary and secondary fermentation stages. Once the target attenuation has been reached, the temperature is allowed to rise (perhaps by 4°C), which permits the yeast to deal more rapidly with the VDKs. Carbonation will be achieved downstream by the direct introduction of gas.

Once the secondary fermentation stage is complete (and the length of this varies considerably between brewers), then the temperature is dropped, ideally to −1°C or −2°C to enable precipitation and sedimentation of materials which would otherwise cause a haze in the beer. The sedimentation of yeast is also promoted in this 'cold conditioning' stage, perhaps with the aid of isinglass finings (Fig. 2.27). These are solutions of collagen derived from the swim bladders of certain species of fish from the South China Seas. Collagen has a net positive charge at the pH of beer, whereas yeast and other particulates have a net negative charge. Opposite charges attracting, the

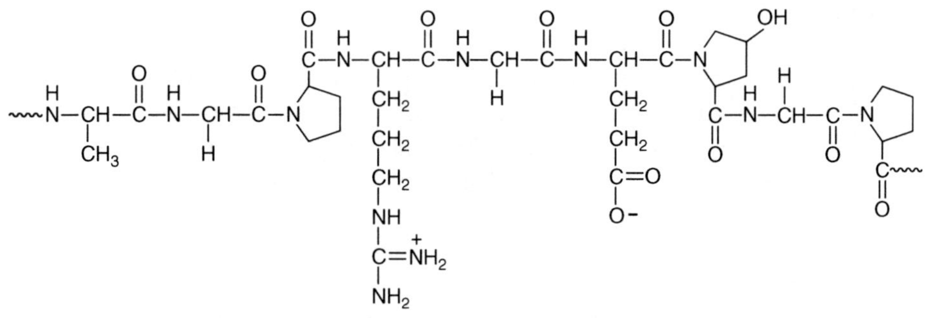

**Fig. 2.27**  A typical repeating structure in the collagen polypeptide chain that, when dissolved in partially degraded forms, represents isinglass. The amino and imino acid residues in this particular sequence are ∼alanyl-prolyl-arginyl-glycyl-glutamyl-hydroxyprolyl-prolyl∼.

isinglass forms a complex with these particles and the resultant large agglomerates sediment readily because of an increase in particle size. Sometimes, the isinglass finings are used alongside 'auxiliary finings' based on silicate, the combination being more effective than isinglass alone. Some brewers centrifuge to aid clarification.

For the most part, fermenters these days are fabricated from stainless steel and will be lagged and feature jackets that allow coolant to be circulated (the heat generated during fermentation is sufficient to effect any necessary warming – so the temperature is regulated by balancing metabolic heat with cooling afforded by the coolant in the jacket, which may be water, glycol or ammonia depending on how much refrigeration is demanded). Modern vessels tend to be enclosed, for microbiological reasons. However, across the world there remain a great many open tanks. Cylindro-conical vessels can have a capacity of up to 13 000 hL and are readily cleaned using CIP operations (see Chapter 1).

Only one company, in New Zealand, practises continuous fermentation. Many brewers nowadays maximise the output by fermenting wort at a higher gravity than necessary to give the target alcohol concentration, before diluting the beer downstream with deaerated water to 'sales gravity' (i.e. the required strength of the beer in package). This is called 'high-gravity brewing'. There are limits to the strength of wort that can be fermented. This is because yeast becomes stressed at high sugar concentrations and when the alcohol level increases beyond a certain point. Brewing is unusual amongst alcohol production industries in that it re-uses yeast for ensuing fermentations. Excluded from this are those beers in which very high alcohol levels are developed (e.g. the barley wines). The yeast is stressed in these conditions and will not be re-usable. This is the reason why wine fermentations, for instance, involve 'one trip' yeast. This is also the reason why, in the production of sweeter fortified wines (see Chapter 4), alcohol is added at the start of fermentation in order to hinder the removal of sugars.

## Filtration

After a period of typically 3 days minimum in 'cold conditioning', the beer is generally filtered. Diverse types of filter are available, perhaps the most common being the plate-and-frame filter which consists of a series of plates in sequence, over each of which a cloth is hung. The beer is mixed with a filter aid – porous particles which both trap particles and prevent the system from clogging. Two major kinds of filter aid are in regular use: kieselguhr and perlite. The former comprises fossils or skeletons of primitive organisms called diatoms. These can be mined and classified to provide grades that differ in their permeability characteristics. Particles of kieselguhr contain pores into which other particles (such as those found in beer) can pass, depending on their size. Perlites are derived from volcanic glasses crushed to form microscopic flat particles. They are better to handle than kieselguhr, but are not as efficient as filter aids. Filtration starts when a pre-coat of filter aid is applied to the filter by cycling a slurry of filter aid through the plates. This pre-coat is generally of quite a coarse grade, whereas the filter-aid (the body feed) which is dosed into the beer during the filtration proper tends to be a finer grade. It is selected according to the particles within the beer that need to be removed. If a beer contains a lot of yeast, but relatively few small particles, then a relatively coarse grade is best. If the converse applies, then a fine grade with smaller pores will be used.

## The stabilisation of beer

Apart from filtration, various other treatments may be applied to beer downstream, all with the aim of enhancing the shelf life of the product. A haze in beer can be due to various materials, but principally it is due to the cross-linking of certain proteins and certain polyphenols. Therefore, if one or both of these materials is removed, then the shelf life is extended. Brewhouse operations are in part designed to precipitate protein–polyphenol complexes. Thus, if these operations are performed efficiently, then much of the job of stabilisation is achieved. Good, vigorous, 'rolling' boils, for instance, will ensure precipitation. Before that, avoidance of the last runnings in the lautering operation will prevent excessive levels of polyphenol entering the wort. The cold conditioning stage also has a major role to play, by chilling out protein–polyphenol complexes, enabling them to be taken out on the filter. Control over oxygen and oxidation is important because it is particularly the oxidised polyphenols that tend to cross-link with proteins. For really long shelf lives, though, and certainly if the beer is being shipped to extremes of climate, additional stabilisation treatments will be necessary. Polyphenols can be removed with PVPP. Protein can be precipitated by adding tannic acid, hydrolysed using papain (the same enzyme from paw paw that is used as meat tenderiser) or, and most commonly, adsorbed on silica hydrogels and silica xerogels.

# Gas control

Final adjustment will now be made to the level of gases in the beer. As we have seen, it is important that the oxygen level in the bright beer is as low as possible. Unfortunately, whenever beer is moved around and processed in a brewery, there is always the risk of oxygen pick-up. For example, oxygen can enter through leaky pumps. A check on oxygen content will be made once the bright beer tank (filtered beer is bright beer) is filled and, if the level is above specification (which most brewers will set at 0.1–0.3 ppm), oxygen will have to be removed. This is achieved by purging the tank with an inert gas, usually nitrogen, from a sinter in the base of the vessel. The level of carbon dioxide in a beer may either need to be increased or decreased. The majority of beers contain between two and three volumes of $CO_2$, whereas most brewery fermentations generate 'naturally' no more than 1.2–1.7 volumes of the gas. The simplest and most usual procedure by which $CO_2$ is introduced is by injection as a flow of bubbles as beer is transferred from the filter to the Bright Beer Tank. If the $CO_2$ content needs to be dropped, this is a more formidable challenge. It may be necessary for beers that are supposed to have a relatively low carbonation and, as for oxygen, this can be achieved by purging. However, concerns about 'bit' production have stimulated the development of gentle membrane-based systems for gas control. Beer is flowed past membranes, made from polypropylene or polytetrafluoroethylene, that are water-hating and therefore do not 'wet-out'. Gases, but not liquids, will pass freely across such membranes, the rate of flux being proportional to the concentration of each individual gas and dependent also on the rate at which the beer flows past the membrane.

# Packaging

The packaging operation is the most expensive stage in the brewery, in terms of raw materials and labour. Beer will be brought into specification in the Bright Beer Tank (sometimes called the Fine Ale Tank or the Package Release Tank). The carbonation level may be higher (e.g. by 0.2 volume) than that specified for the beer in package, to allow for losses during filling.

Although beer is relatively resistant to spoilage, it is by no means entirely incapable of supporting the growth of micro-organisms. For this reason, most beers are treated to eliminate any residual brewing yeast or infecting wild yeasts and bacteria before or during packaging. This can be achieved in one of two ways: pasteurisation or sterile filtration. Pasteurisation can take one of two forms in the brewery: flash pasteurisation for beer pre-package and tunnel pasteurisation for beer in can or bottle. The principle in either case, of course, is that heat kills micro-organisms. One PU is defined as exposure for 1 min at 60°C. The higher the temperature, the more rapidly the micro-organisms are destroyed. A 7°C rise in temperature leads to a ten-fold increase in the rate of cell death. The pasteurisation time required to kill organisms at different temperatures can be read off from a plot. Typically, a brewer might use 5–20 PU – but higher 'doses' may be used for some beers, for example, low alcohol beers which are more susceptible to infection. In flash pasteurisation, the beer flows through a heat exchanger (essentially like a wort cooler acting in reverse), which raises the temperature typically to 72°C. Residence times of between 30 and 60 s at this temperature are sufficient to kill off virtually all microbes. Ideally there will not be many of these to remove: good brewers will ensure low loadings of micro-organisms by attention to hygiene throughout the process and ensuring that the prior filtration operation is efficient. Tunnel pasteurisers comprise large heated chambers through which cans or glass bottles are conveyed over a period of minutes, as opposed to the seconds employed in a flash pasteuriser. Accordingly, temperatures in a tunnel pasteuriser are lower, typically 60°C for a residence time of 10–20 min. An increasingly popular mechanism for removing micro-organisms is to filter them out by passing the beer through a fine mesh filter. The rationale for selecting this procedure rather than pasteurisation is as much for marketing reasons as for any technical advantage it presents: many brands of beer these days are being sold on a claim of not being heat-treated, and therefore free from any 'cooking'. In fact, provided the oxygen level is very low, modest heating of beer does not have a major impact on the flavour of many beers, although those products with relatively subtle, lighter flavour will obviously display 'cooked' notes more readily than will beers that have a more complex flavour character. The sterile filter must be located downstream from the filter that is used to separate solids from the beer. Sterile filters may be of several types, a common variant incorporating a membrane formed from polypropylene or polytetrafluoroethylene and with pores of between 0.45 and 0.8 μm.

### Filling bottles and cans

Bottles entering the brewery's packaging hall are first washed and, if they are returnable bottles (i.e. they have been used previously to hold beer), they will need a much more robust cleaning and sterilisation, inside and out, involving soaking and jetting with hot caustic detergent and thorough rinsing with water. The beer coming from the Bright Beer Tanks is transferred to a bowl at the heart of the filling machine. Bottle fillers are machines based on a rotary carousel principle. They have a series of filling heads: the more the heads, the greater the capacity of the filler. The bottles enter on a conveyor and, sequentially, each is raised into position beneath the next vacant filler head, each of which comprises a filler tube. An air-tight seal is made and, in modern fillers, a specific air evacuation stage starts the filling sequence. The bottle is counter-pressured with carbon dioxide, before the beer is allowed to flow into the bottle by gravity from the bowl. The machine will have been adjusted so that the correct volume of beer is introduced into the vessel. Once filled, the 'top' pressure on the bottle is relieved, and the bottle is released from its filling head. It passes rapidly to the machine that will crimp on the crown cork but, en route, the bottle will have been either tapped or its contents 'jetted' with a minuscule amount of sterile water in order to fob the contents and drive off any air from the space in the bottle between the surface of the beer and the neck (the 'headspace'). Next stop is the tunnel pasteuriser if the beer is to be pasteurised after filling, but if sterile filtration is used, the filler and capper are likely to be enclosed in a sterile room. The bottles now head off for labelling, secondary packaging and warehousing.

Putting beer into cans has much in common with bottling. It is the container, of course, that is very different – and definitely one trip. Cans may be of aluminium or stainless steel, which will have an internal lacquer to protect the beer from the metal surface and vice versa. Cans arrive in the canning hall on vast trays, all pre-printed and instantly recognisable. They are inverted, washed and sprayed, prior to filling in a manner very similar to the bottles. Once filled, the lid is fitted to the can basically by folding the two pieces of metal together to make a secure seam past which neither beer nor gas can pass.

### Filling kegs

Kegs are manufactured from either aluminium or stainless steel. They are containers generally of 100 L or less, containing a central spear. Kegs, of course, are multi-trip devices. On return to the brewery from an 'outlet', they are washed externally before transfer to the multi-head machine in which successive heads are responsible for their washing, sterilising and filling. Generally they will be inverted as this takes place. The cleaning involves high-pressure spraying of the entire internal surface of the vessel with water at approximately 70°C. After about 10 s, the keg passes to the steaming stage, the temperature reaching 105°C over a period of perhaps half a minute. Then the keg goes to the filling head, where a brief purge with carbon dioxide precedes the introduction of beer, which may take a couple of minutes. The discharged keg is weighed to ensure that it contains the correct quantity of beer and is labelled and palleted before warehousing.

## The quality of beer

### Flavour

The flavour of beer can be split into three separate components: taste, smell (aroma) and texture (mouthfeel).

There are only four proper tastes: sweet, sour, salt and bitter. They are detected on the tongue. A related sense is the tingle associated with high levels of carbonation in a drink: this is due to the triggering of the trigeminal nerve by carbon dioxide. This nerve responds to mild irritants, such as carbonation and capsaicin (a substance largely responsible for the 'pain delivery' of spices and peppers).

Carbon dioxide is also relevant insofar as its level influences the extent to which volatile molecules will be delivered via the foam and into the headspace above the beer in a glass.

The sweetness of a beer is due, of course, to its level of sugars, either those that have survived fermentation or those introduced as primings.

The principal contributors to sourness in beer are the organic acids that are produced by yeast during fermentation. These lower the pH: it is the $H^+$ ion imparted by acidic solutions that causes the sour character to be perceived on the palate. Most beers have a pH between 3.9 and 4.6.

Saltiness in beer is afforded by sodium and potassium, while of the anions present in beer, chloride and sulphate are of particular importance. Chloride is said to contribute a mellowing and fullness to a palate, while sulphate is felt to elevate the dryness of beer.

Perhaps the most important taste in beer is bitterness, primarily imparted by the iso-$\alpha$-acids derived from the hop resins.

Many people believe that they can taste other notes on a beer. In fact they are detecting them with the nose, the confusion arising because there is a continuum between the back of the throat and the nasal passages. The smell (or aroma) of a beer is a complex distillation of the contribution of a great many individual molecules. No beer is so simple as to have its 'nose' determined by one or even a very few substances. The perceived character is a balance between positive and negative flavour notes, each of which may be a consequence of one or a combination of many compounds of different chemical classes. The 'flavour threshold' is the lowest concentration of a substance which is detectable in beer.

The substances that contribute to the aroma of beer are diverse. They are derived from malt and hops and by yeast activity (leaving aside for the moment the contribution of contaminating microbes). In turn there are interactions between these sources, insofar as yeast converts one flavour constituent from malt or hops into a different one, for example.

Various alcohols influence the flavour of beer (Table 2.5), by far the most important of which is ethanol, which is present in most beers at levels at least 350-fold higher than any other alcohol. Ethanol contributes directly to the

**Table 2.5**  Some alcohols in beer.

| Alcohol | Flavour threshold (mg L$^{-1}$) | Perceived character |
|---|---|---|
| Ethanol | 14000 | Alcoholic |
| Propan-1-ol | 800 | Alcoholic |
| Butan-2-ol | 16 | Alcoholic |
| Iso-amyl alcohol | 50 | Alcohol, banana, vinous |
| Tyrosol | 200 | Bitter |
| Phenylethanol | 40–100 | Roses, perfume |

**Table 2.6** Some esters in beer.

| Ester | Flavour threshold (mg L$^{-1}$) | Perceived character |
|---|---|---|
| Ethyl acetate | 33 | Solventy, fruity, sweet |
| Iso-amyl acetate | 1.0 | Banana |
| Ethyl octanoate | 0.9 | Apples, sweet, fruity |
| Phenylethyl acetate | 3.8 | Roses, honey, apple |

flavour of beer, registering a warming character. It also influences the flavour contribution of other volatile substances in beer. Because it is quantitatively third only to water and carbon dioxide as the main component of beer, it is not surprising that it moderates the flavour impact of other substances. It does this by affecting the vapour pressure of other molecules (i.e. their relative tendency to remain in beer or to migrate to the headspace of the beer). The higher alcohols in beer are important as the immediate precursors of the esters, which are proportionately more flavour active (see Table 2.6). And so it is important to be able to regulate the levels of the higher alcohols produced by yeast if ester levels are also to be controlled.

The higher alcohols are produced during fermentation by two routes: catabolic and anabolic. In the catabolic route, yeast amino acids taken up from the wort by yeast are transaminated to $\alpha$-keto-acids, which are decarboxylated and reduced to alcohols:

$$RCH(NH_2)COOH + R^1COCOOH \rightarrow RCOCOOH + R^1CH(NH_2)COOH \tag{2.1}$$

$$RCOCOOH \rightarrow RCHO + CO_2 \tag{2.2}$$

$$RCHO + NADH + H^+ \rightarrow RCH_2OH + NAD^+ \tag{2.3}$$

The anabolic route starts with pyruvate (the end point of the EMP pathway proper), the higher alcohols being 'side shoots' from the synthesis of the amino acids valine and leucine (Fig. 2.28). The penultimate stage in the production of all amino acids is the formation of the relevant keto acid which is transaminated to the amino acid. Should there be conditions where the keto acids accumulate, they are then decarboxylated and reduced to the equivalent alcohol. Essentially, therefore, the only difference between the pathways is the origin of the keto acid: either the transamination product of an amino acid assimilated by the yeast from its growth medium or synthesised *de novo* from pyruvate.

In view of the above, it is not surprising that the levels of FAN in wort influence the levels of higher alcohols formed. Higher alcohol production is increased at both excessively high and insufficiently low levels of assimilable nitrogen available to the yeast from wort. If levels of assimilable N are low, then yeast growth is limited and there is a high incidence of the anabolic pathway. If levels of N are high, then the amino acids feedback to inhibit further synthesis of them and therefore the anabolic pathway becomes less important. However, there is a greater tendency for the catabolic pathway to 'kick in'.

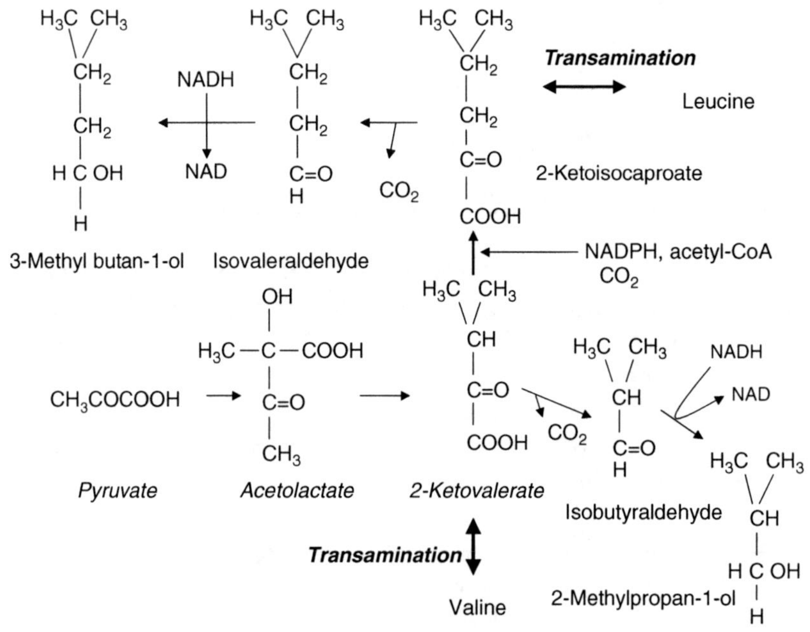

**Fig. 2.28**  The anabolic route to higher alcohols in yeast. Note: Fig. 2.29 shows how acetolactate is derived from pyruvate.

Even more important than FAN levels, though, is the yeast strain, with ale strains producing more of these compounds than lager strains. Fermentations at higher temperatures increase higher alcohol production. Conditions favouring increased yeast growth (e.g. excessive aeration or oxygenation) promote higher alcohol formation, but this can be countered by application of a top pressure on the fermenter. The reasons why increased pressure has this effect are unclear, but it has been suggested that it may for some reason be due to an accumulation of carbon dioxide. Whatever the reason, it is pertinent to mention that beer produced in different sizes and shapes of vessel, displaying different hydrostatic pressures, do produce higher alcohols (and thereof esters) to different extents. This can be a problem for product matching between breweries (e.g. in franchise brewing operations).

Various esters may make a contribution to the flavour of beer (Table 2.6). The esters are produced from their equivalent alcohols (ROH), through catalysis by the enzyme alcohol acetyl transferase (AAT), with acetyl-coenzyme A being the donor of the acetate group:

$$ROH + CH_3COSCoA \rightarrow CH_3COOR + CoASH$$

Clearly the amount of ester produced will depend *inter alia* on the levels of acetyl-CoA, of alcohol and of AAT. Esters are formed under conditions when the acetyl-CoA is not required as the prime building block for the synthesis of key cell components. In particular, acetyl-CoA is the starting point for the synthesis of lipids, which the cell requires for its membranes. Thus, factors promoting yeast production (e.g. high levels of aeration/oxygenation) *lower* ester production, and vice versa.

However, perhaps the most significant factor influencing the extent of ester production is yeast strain, some strains being more predisposed to generating esters than others. This may relate to the amount of AAT that they contain. The factors that dictate the level of this enzyme present in a given yeast strain are not fully elucidated, but it does seem to be present in raised quantities when the yeast encounters high-gravity wort, and this may explain the disproportionate extent of ester synthesis under these conditions.

Whereas the esters and higher alcohols can make positive contributions to the flavour of beer, few beers (with the possible exception of some ales) are helped by the presence of VDKs, diacetyl and (less importantly) pentanedione (Table 2.7). Elimination of VDKs from beer depends on the fermentation process being well-run. These substances are offshoots of the pathways by which yeast produces the amino acids valine and isoleucine (and therefore there is a relationship to the anabolic pathway of higher alcohol production).

The pathway for diacetyl production is shown in Fig. 2.29 because it is more significant (with respect to diacetyl being present at higher levels and with a lower flavour threshold). The precursor molecules leak out of the yeast and break down spontaneously to form VDKs. Happily, the yeast can mop up the VDK, *provided* it remains in contact with the beer and is in good condition.

Reductases in the yeast reduce diacetyl successively to acetoin and 2,3-butanediol, both of which have much higher flavour thresholds than diacetyl.

**Table 2.7** VDKs in beer.

| VDK | Flavour threshold (mg L$^{-1}$) | Perceived character |
| --- | --- | --- |
| Diacetyl | 0.1 | Butterscotch |
| Pentanedione | 0.9 | Honey |

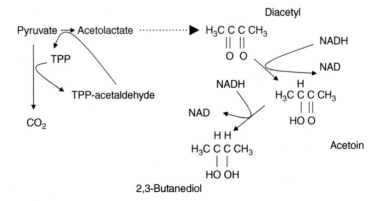

**Fig. 2.29** The production and elimination of diacetyl by yeast.

**Table 2.8** Some sulphur-containing substances in beer.

| S-containing compound | Flavour threshold (mg $L^{-1}$) | Perceived character |
|---|---|---|
| Hydrogen sulphide | 0.005 | Rotten eggs |
| Sulphur dioxide | 25 | Burnt matches |
| Methanethiol | 0.002 | Drains |
| Ethanethiol | 0.002 | Putrefaction |
| Propanethiol | 0.0015 | Onion |
| Dimethyl sulphide | 0.03 | Sweetcorn |
| Dimethyl disulphide | 0.0075 | Rotting vegetables |
| Dimethyl trisulphide | 0.0001 | Rotting vegetables, onion |
| Methyl thioacetate | 0.05 | Cooked cabbage |
| Diethyl sulphide | 0.0012 | Cooked vegetables, garlic |
| Methional | 0.25 | Cooked potato |
| 3-Methyl-2-butene-1-thiol | 0.000004–0.001 | Lightstruck, skunky |
| 2-Furfurylmercaptan | | Rubber |

Many brewers allow a temperature rise at the end of fermentation to facilitate more rapid removal of VDKs. Others introduce a small amount of freshly fermenting wort later on as an inoculum of healthy yeast (a process known as *Krausening*). Persistent high diacetyl levels in a brewery's production may be indicative of an infection by Pediococcus or Lactobacillus bacteria. If the ratio of diacetyl to pentanedione is disproportionately high, then this indicates that there is an infection problem.

In many ways the most complex flavour characters in beer are due to the sulphur-containing compounds. There are many of these in beer (Table 2.8) and they make various contributions. Thus, many ales have a deliberate hydrogen sulphide character, not too much, but just enough to give a nice 'eggy' nose. Lagers tend to have a more complex sulphury character. Some lagers are relatively devoid of any sulphury nose. Others, though, have a distinct DMS character, while some have characters ranging from cabbagy to burnt rubber. This range of characteristics renders substantial complexity to the control of sulphury flavours.

All of the DMS in a lager ultimately originates from a precursor, S-methylmethionine (SMM), produced during the germination of barley (Fig. 2.30). SMM is heat sensitive and is broken down rapidly whenever the temperature gets above about 80°C in the process. Thus, SMM levels are lower in the more intensely kilned ale malts and, as a result, DMS is a character more associated with lagers. SMM leaches into wort during mashing and is further degraded during boiling and in the whirlpool. If the boil is vigorous, most of the SMM is converted to DMS and this is driven off. In the whirlpool, though, conditions are gentler and any SMM surviving the boil will be broken down to DMS but the latter tends to stay in the wort. Brewers seeking to retain some DMS in their beer will specify a finite level of SMM in their malt and will manipulate the boil and whirlpool stages in order to deliver a certain level of DMS into the pitching wort. During fermentation, much DMS will be lost with the gases, so the level of DMS required in the wort will be somewhat

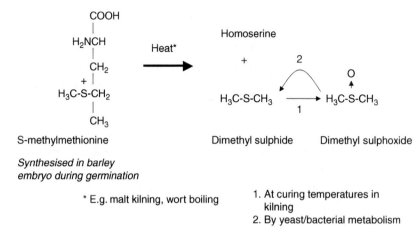

**Fig. 2.30**   The origin of DMS in beer.

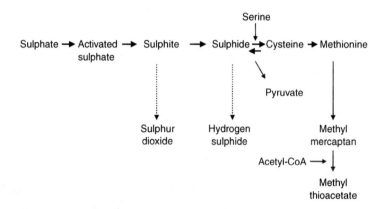

**Fig. 2.31**   The origins of other sulphur-containing volatiles in beer (see also Fig. 1.17 in Chapter 1).

higher than that specified for the beer. There is another complication, insofar as some of the SMM is converted into a third substance, DMSO, during kilning. This is not heat-labile but is water-soluble. It gets into wort at quite high levels and some yeast strains are quite adept at converting it to DMS.

Hydrogen sulphide ($H_2S$) can also be produced by more than one pathway in yeast. It may be formed by the breakdown of amino acids such as cysteine or peptides like glutathione, or by the reduction of inorganic sources such as sulphate and sulphite (Fig. 2.31). Once again, yeast strain has a major effect on the levels of $H_2S$ that are produced during fermentation. For all strains, more $H_2S$ will be present in green beer if the yeast is in poor condition, because a vigorous fermentation is needed to purge $H_2S$. Any other factor that hinders fermentation (e.g. a lack of zinc or vitamins) will also lead to an exaggeration of $H_2S$ levels in beer. Furthermore, $H_2S$ is a product of yeast autolysis, which will be more prevalent in unhealthy yeast.

When the bitter iso-$\alpha$-acids are exposed to light, they break down, react with sulphur sources in the beer and form a substance called 3-methyl-2-butene-1-thiol (MBT), which has an intense skunky character and is detectable at extremely low concentrations. There are two ways of protecting beer from this: do not expose beer to light or else bitter using chemically modified bitter extracts, the reduced iso-$\alpha$-acids.

The addition of hops during beer production not only contributes much of the resulting bitterness, but also imparts a unique so-called 'hoppy' aroma. This attribute comes from the complex volatile oil fraction of hops. Most of the component substances do not survive the brewing process intact and are chemically transformed into as yet poorly defined compounds. Certainly, there does not appear to be one compound solely responsible for hop aroma in beer, although several groups (e.g. sesquiterpene epoxides, cyclic ethers and furanones) have been strongly implicated.

The point at which hops are added during beer production determines the resulting flavour that they impart. The practice of adding aroma hops close to the end of boiling (late hopping) still results in the substantial evaporation of volatile material, but of the little that remains, much is transformed into other species (e.g. the hop oil component humulene can be converted to the more flavour-active humulene epoxide). Further changes then occur during fermentation, such as the transesterification of methyl esters to their ethyl counterparts. The resultant late hop flavour is rather floral in character and is generally an attribute more associated with lager beers.

In a generally distinct practice, hops may be added to the beer right at the end of production. This process of dry hopping gives certain ales their characteristic aroma. The hop oil components contributed to beer by this process are very different to those from late hopping, with mono- and sesquiterpenes surviving generally unchanged in the beer.

Malty character in beer is due in part at least to isovaleraldehyde, which is formed by a reaction between one of the amino acids (leucine) and reductones in the malt. The toffee and caramel flavours in crystal malts and the roasted, coffee-like notes found in darker malts are due to various complex components generated from amino acids and sugars that cross-react during kilning – the Maillard reaction (see Chapter 1).

Acetaldehyde, which is the immediate precursor of ethanol in yeast, has a flavour threshold of between 5 and 50 mg L$^{-1}$ and imparts a 'green apples' flavour to beer. High levels should not survive into beer in successful fermentations, because yeast will efficiently convert the acetaldehyde into ethanol. If levels are persistently high, then this is an indication of premature yeast separation, poor yeast quality or a Zymomonas infection.

The *short-chain fatty acids* (Table 2.9) are made by yeast as intermediates in the synthesis of the lipid membrane components. For this reason, the control of these acids is exactly analogous to that of the esters (discussed earlier): if

**Table 2.9** Some short-chain fatty acids in beer.

| Fatty acid | Flavour threshold ($mg\,L^{-1}$) | Perceived character |
| --- | --- | --- |
| Acetic | 175 | Vinegar |
| Propionic | 150 | Acidic, milky |
| Butyric | 2.2 | Cheesy |
| 3-Methyl butyric | 1.5 | Sweaty |
| Hexanoic | 8 | Vegetable oil |
| Octanoic | 15 | Goaty |
| Phenyl acetic | 2.5 | Honey |

yeast needs to make fewer lipids (under conditions where it needs to grow less), then short-chain fatty acids will accumulate.

Some beers (e.g. some wheat beers) feature a phenolic or clove-like character. This is due to molecules such as 4-vinylguaiacol (4-VG), which is produced by certain *Saccharomyces* species, including *Saccharomyces diastaticus*. Its unwanted presence in a beer is an indication of a wild yeast infection. 4-VG is produced by the decarboxylation of ferulic acid by an enzyme that is present in *S. diastaticus* and other wild yeasts, but not in brewing strains other than a few specific strains of *S. cerevisiae*, namely the ones prized in Bavaria for their use in wheat beer manufacture.

A further undesirable note is a metallic character which, if present in beer, is most likely to be due to the presence of high levels (>0.3 ppm) of iron. One known cause is the leaching of the metal from filter aid.

The flavour of beer changes with time. There is a decrease in bitterness (due to the progressive loss of the iso-$\alpha$-acids), an increase in perceived sweetness and toffee character and a development of a cardboard note. It is the cardboard note that most brewers worry about in connection with the shelf life of their products. Cardboard is due to a range of carbonyl compounds, which may originate in various precursors, including unsaturated fatty acids, higher alcohols, amino acids and the bitter substances. Most importantly, their formation is a result of oxidation, hence the importance of minimising oxygen levels in beer and, perhaps, further upstream.

Any drinker who has ordered a beer containing nitrogen gas will appreciate that one can talk of the mouthfeel and texture of beer. $N_2$ not only imparts a tight, creamy head to a beer, but it also gives rise to a creamy texture. More specifically, the partial replacement of carbon dioxide with nitrogen gas suppresses several beer flavour attributes, such as astringency, bitterness, hop aroma as well as the reduction in the carbon dioxide 'tingle'. Other components of beer, such as the astringent polyphenols, may also play a part. Physical properties, such as viscosity, are influenced by residual carbohydrate in the beer and might also contribute to the overall mouthfeel of a product. It is thought that turbulent flow of liquids between the tongue and the roof of the mouth results in increased *perceived* viscosity and therefore enhanced mouthfeel.

## Foam

A point of difference between beer and other alcoholic beverages is its possession of stable foam. This is due to the presence of hydrophobic (amphipathic) polypeptides, derived from cereal, that cross-link with the bitter iso-$\alpha$-acids in the bubble walls to counter the forces of surface tension that tend to lead to foam collapse.

## Gushing

Foaming can be taken to excess, in which case the problem which manifests itself in small pack is 'gushing', that is, the spontaneous generation of foam on opening a package of beer. This is due to the presence of nucleation sites in beer that cause the dramatic discharging of carbon dioxide from solution. These nucleation sites may be particles of materials like oxalate or filter aid, but most commonly gushing is caused by intensely hydrophobic peptides that are produced from Fusarium that can infect barley unless precautions are taken.

## Spoilage of beer

Compared with most other foods and beverages beer is relatively resistant to infection. There are several reasons for this, namely the presence of ethanol, a low pH, the relative shortage of nutrients (sugars, amino acids), the anaerobic conditions and the presence of antimicrobial agents, notably the iso-$\alpha$-acids.

The most problematic Gram-positive bacteria are lactic acid bacteria belonging to the genera *Lactobacillus* and *Pediococcus*. At least ten species of lactobacillus spoil beer. They tolerate the acidic conditions. Some species (e.g. *Lactobacillus brevis* and *Lactobacillus plantarum*) grow quickly during fermentation, conditioning and storage, while others (e.g. *Lactobacillus lindner*) grow relatively slowly. Spoilage with lactobacilli is especially problematic during the conditioning of beer and after packaging, resulting in a silky turbidity and off-flavours. Pediococci are homofermentative. Six species have been identified, the most important being *Pediococcus damnosus*. Such infection generates lactic acid and diacetyl. The production of polysaccharide capsules can cause ropiness in beer.

Many Gram-positive bacteria are killed by iso-$\alpha$-acids. These agents probably disrupt nutrient transport across the membrane of the bacteria, but only when they are present in their protonated forms (i.e. at low pH). This is one of the reasons why a beer at pH 4.0 will be more resistant to infection than one at pH 4.5. Some Gram positives are resistant to iso-$\alpha$-acids and most Gram negatives are.

Important Gram-negative bacteria include the acetic acid bacteria (*Acetobacter*, *Gluconobacter*); Enterobacteriaceae (*Escherichia*, *Aerobacter*,

*Klebsiella, Citrobacter, Obesumbacterium*); *Zymomonas, Pectinatus* and *Megasphaera*. Acetic acid bacteria produce a vinegary flavour in beer and a ropy slime. It is most often found in draft beer, where there is a relatively aerobic environment close to the beer, for example, in partly emptied containers. Enterobacteriaceae are aerobic and cannot grow in the presence of ethanol. They are a threat in wort and early in fermentation and they produce cabbagy/vegetable/eggy aromas. *Zymomonas* is a problem with primed beers (it uses invert sugar or glucose, but cannot use maltose). Although it has a metabolism reminiscent of *Saccharomyces* (it's actually used to produce alcoholic beverages in some countries), it does tend to produce large amounts of acetaldehyde.

**Table 2.10**  Major beer styles.

| Style | Origin | Notes |
| --- | --- | --- |
| *(a) Ales and stouts* | | |
| Bitter (pale) ale | England | Dry hop, bitter, estery, malty, low carbonation (on draught), copper colour |
| India Pale Ale | England | Similar, but substantially more bitter |
| Alt (n.b. Alt means 'old') | Germany | Estery, bitter, copper colour |
| Mild (brown) ale | England | Darker than pale ale, malty, slightly sweeter, lower in alcohol |
| Porter | England | Dark brown/black, less 'roast' character than stout, malty |
| Stout | Ireland | Black, roast, coffee-like, bitter |
| Sweet stout | England | Caramel-like, brown, full-bodied |
| Imperial stout | England | Brown/black, malty, alcoholic |
| Barley wine | England | Tawny/brown, malty, alcoholic, warming |
| Kölsch | Germany | Straw/golden colour, caramel-like, medium bitterness, low hop aroma |
| Weizenbier (wheat beer) | Germany | Hefeweissens retain yeast (i.e. turbid). Kristalweissens are filtered. Very fruity, clove-like, high carbonation |
| Lambic | Belgium | Estery, sour, 'wet horse-blanket', turbid. Lambic may be mixed with cherry (kriek), peach (peche), raspberry (framboise), etc. Old lambic blended with freshly fermenting lambic = gueuze |
| Saison | Belgium | Golden, fruity, phenolic, mildly hoppy |
| *(b) Lagers* | | |
| Pilsener | Czech | Golden/amber, malty, late hop aroma |
| Bock | Germany | Golden/brown, malty, moderately bitter |
| Helles | Germany | Straw/golden, low bitterness, malty, sulphury |
| Märzen (meaning 'March' for when traditionally brewed) | Germany | Diverse colours, sweet malt flavour, crisp bitterness |
| Vienna | Austro-Hungaria | Red-brown, malty, toasty, crisply bitter |
| Dunkel | Germany | Brown, malty, roast-chocolate |
| Schwarzbier | Germany | Brown/black, roast malt, bitter |
| Rauchbier | Germany | Smokey |
| Malt liquor | USA | Pale colour, alcoholic, slightly sweet, low bitterness |

A wild yeast is any yeast other than the culture yeast used for a given beer. As well as *Saccharomyces*, wild yeast may be *Brettanomyces, Candida, Debaromyces, Hansenula, Kloeckera, Pichia, Rhodotorula, Torulaspora* or *Zygosaccharomyces*. If the contaminating yeast is another brewing yeast, then the risk is a shift in performance to that associated with the 'foreign' yeast (i.e. you will not get the expected beer). If the contaminant is another type of yeast, the risk is off-flavour production (e.g. clove-like flavours produced by decarboxylation of ferulic acid) or a problem like over-attenuation as might be caused by a diastatic organism such as *S. diastaticus*.

## Beer styles

An indication of the complexity of beer styles available worldwide will be gleaned from Table 2.10. In relation to the immediately foregoing discussion, we might note the lambic and gueuze products of Belgium, whose production depends not only on Saccharomyces species, but also *inter alia* Pediococcus, Lactobacillus, Brettanomyces, Candida, Hansenula and Pichia.

## Bibliography

Bamforth, C.W. (2003) *Beer: Tap into the Art and Science of Brewing*, 2nd edn. New York: Oxford University Press.

Baxter, E.D. & Hughes, P.S. (2001) *Beer: Quality, Safety and Nutritional Aspects.* London: Royal Society of Chemistry.

Boulton, C. & Quain, D. (2001) *Brewing Yeast and Fermentation.* Oxford: Blackwell Publishing.

Briggs, D.E. (1998) *Malts and Malting.* London: Blackie.

Briggs, D.E., Boulton, C.A., Brookes, P.A. & Stevens, R. (2004) *Brewing: Science and Practice.* Cambridge: Woodhead.

MacDonald, J., Reeve, P.T.V., Ruddlesden, J.D. & White, F.H. (1984) Current approaches to brewery fermentations. In *Progress in Industrial Microbiology*, vol. 19 (ed. M.E. Bushell), pp. 47–198. Amsterdam: Elsevier.

MacGregor, A.W. & Bhatti, R.S., eds (1993) *Barley: Chemistry and Technology.* St Paul, MN: American Association of Cereal Chemists.

Neve, R.A. (1991) *Hops.* London: Chapman & Hall.

# Chapter 3
# Wine

The *Merriam-Webster's Dictionary* defines wine as *the usually fermented juice of a plant product (as a fruit) used as a beverage*. While in rural communities in countries such as Great Britain wines have from time immemorial been produced from all manner of plant materials (and not only fruits), I restrict discussion in the present chapter to the products of commercial entities furnishing wines based on the grape (Fig. 3.1).

## Grapes

The importance of sound viticulture as a precursor to wines of excellence is unequivocally accepted as a truth in wine making companies worldwide. More so than for beer is the belief held that it is not possible to make an excellent product unless there is similar excellence in the source of fermentable carbohydrate. Most wineries tend to grow their own grapes or buy them from nearby vineyards.

The ideal climate for growing wine grapes is where there is no summer rain, it is hot or at least warm during the day, there are cool nights and little risk of frost damage. The great grape-growing and wine regions are listed in Table 3.1. A benchmark figure for the yield of wine from one metric ton of grapes would be around 140–160 gal. As red grapes are fermented on the skins and therefore are less demanding in the pressing stage, the yield is some 20% higher than for whites.

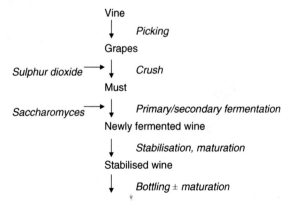

**Fig. 3.1**   An overview of wine making.

**Table 3.1**   Major wine grape-growing regions (1998).

| Country | Wine production (thousand litres) | Grape production (thousand tons) |
|---|---|---|
| Italy | 5.42 | 9.21 |
| France | 5.27 | 6.88 |
| Spain | 3.03 | 4.88 |
| United States | 2.05 | 5.36 |
| Argentina | 1.27 | 2.00 |
| Germany | 1.08 | 1.41 |
| South Africa | 0.82 | 1.30 |
| Australia | 0.74 | |
| Chile | 0.55 | |
| Romania | 0.50 | |

Data derived from Dutruc-Rosset, G. (2000).

**Fig. 3.2**   Scion buds grafted on to the rootstock. Courtesy of E & J Gallo.

Wine grapes belong to the genus *Vitis*. Within the genus, the main species are *vinifera* (by far the most important), *lubruscana* and *rotundifolia*. Commercial vines tend to be *Vitis vinifera* grafter onto rootstocks of the other *Vitis* species. Of course within the species is a diversity of varieties (cultivars) – for example, *V. vinifera* var. Cabernet Sauvignon.

It takes approximately 4–5 years from the first planting to yield the first truly good crop of grapes. The scion (top) of the vine and the rootstock to which it is grafted (Fig. 3.2) must be selected on the basis of compatibility, one with the other and the combination with the local soil and climate. Other key issues that come to bear in viticulture are the availability of sunlight, depth of the soil, its nutrient and moisture content and how readily it drains.

**Table 3.2** Some varieties of grape.

| Type | Example | Comments |
|------|---------|----------|
| *White cultivars* | | |
| Messiles | Sauvignon blanc | Bordeaux. Green pepper and herbaceous notes |
| Muscats | Muscat blanc | Raisin notes. Prone to oxidation, so often made into dessert wines |
| Noiriens | Chardonnay | Widespread use globally; use in champagne production. Wines have apple, melon, peach notes |
| Parellada | | Catalonia. Apple/citrus notes |
| Rhenans | Gewurztraminer | Cooler European regions. Lychee characters |
| Riesling | | German origin. Rose and pine notes |
| Viura | | Rioja. Butterscotch and banana |
| *Red cultivars* | | |
| Carmenets | Cabernet Sauvignon Merlot | Bordeaux. Tannic. Blackcurrant aroma Lighter in character |
| Nebbiolo | | Italy. Acid, tannic. Truffle, tar and violet notes |
| Noiriens | Pinot Noir | Beet, cherry, peppermint notes when optimal |
| Sangiovese | | Chianti. Cherries, violets, liquorice |
| Serines | Syrah | France (n.b. Shiraz in Spain). Tannic, peppery aromas |
| Tempranillo | | Spain, especially Rioja. Also grown in Argentina. Jam, citrus, incense notes |
| Zinfandel | | California. Also used for light blush wines |

Some regions are especially susceptible to diseases such as Pierce's disease and phylloxera (an insect that attacks rootstock and which is prevalent, for instance, in the Eastern United States but now also in California).

Vines should go dormant in order to survive cold winters. Cool autumn conditions with light or medium frosts allow the vine to store enough carbohydrate for good growth in the ensuing spring. There may be 500–600 or more vines per acre. New vines are trained up individual stakes in the first growing season. Only one shoot is trained in each instance with the others being pinched off. Pruning of vines takes place in winter months after the vines have proceeded to dormancy and the canes have hardened and turned brown.

It is important to match grape variety to the location and to the style of wine. A variety may develop certain characteristics earlier depending on how warm the growing region is. Accordingly, when that grape achieves full maturity, it may have lost some of that character. Table 3.2 summarises varietal issues.

There is some understanding (though far from complete) of the chemistry involved in varietal differences. For instance, methyl anthranilate is found in Lambrusca, 2-methoxy-3-isobutylpyrazine in Cabernet Sauvignon, damascenone in Chardonnay (Fig. 3.3). For muscats there are terpenes such as linalool and geraniol and there are terpenols in White Riesling. Some of these are found in the form of complexes with sugars known as glycosides (Fig. 3.4). Yeast produces enzymes called glycosidases that sever the link between the flavour-active molecule and the sugar over time, illustrating the time dependence of flavour development in this type of wine.

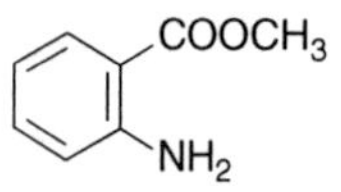

Methyl anthranilate

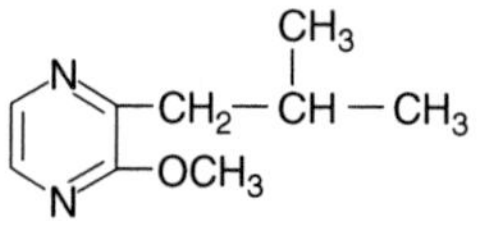

2-Methoxy-3-isobutylpyrazine

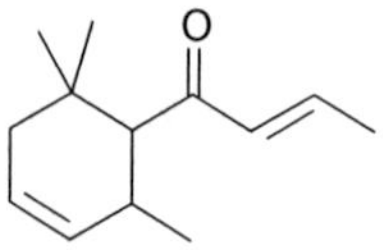

Damascenone

**Fig. 3.3**   Some compounds responsible for varietal differences in wines.

Sugar – aglycone   → (*Glycosidase*)   Sugar + aglycone

Glycoside

*Examples of aglycones*: terpenes and terpenols

For example, geraniol

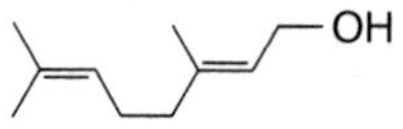

**Fig. 3.4**   Glycosides and glycosidases.

Unless a soil is extremely acidic or alkaline or suffers from deficient drainage, the soil type *per se* is unlikely to be a major issue with regard to grape quality. Any deficiencies in nitrogen level will need to be corrected by adding N, avoiding excess so as not to promote wasteful growth of non-grape tissue or increase the risk of spoilage and development of ethyl carbamate.

The local climate also influences the susceptibility of the vines to infestation. If there are rains in summer months or if the vineyard is afforded excessive irrigation, there is an increased risk of powdery or downy mildew. Excess water uptake by grapes can also cause berries to swell and burst, which in turn enables rot and mould growth. Over-watering leads to excessive cane growth and delays the maturation of the fruit. In regions where infestation and infection are a particular problem, it is likely that some form of chemical treatment will be necessary. *Botyritis cinerea* is common where summer rains are prevalent. Winemakers refer to it as grey mould when regarded in an unfavourable light but as 'noble rot' when deemed desirable. The contamination leads to oxidation of sugars and depletion of nitrogen, as well as reduction of certain desirable flavours. However, the character of certain wines depends on this infection, for example, the Sauternes from France.

Of particular alarm in some grape-growing regions is Pierce's disease. This is caused by the bacterium *Xylella fastidiosa* and is spread by an insect known as the glassy-winged sharpshooter. It is prevalent in North and Central America, and is of annual concern in some Californian vineyards. It appears to be restricted to regions with mild winters. The sharpshooter feeds on xylem sap and transmits bacteria to the healthy plant. The water-conducting system is blocked and there is a drying or 'scorching' of leaves, followed by the wilting of grape clusters.

Harvesting of grapes is usually in the period from August through September and October. The time of harvesting has a significant role to play in determining the sweetness/acid balance of grapes. Grapes grown in warm climates tend to lose their acidity more rapidly than do those in cooler environs. This loss of acidity is primarily due to respiratory removal of malic acid during maturation. The other key acid, tartaric, is less likely to change in level.

Ripe fruit should be picked immediately before it is to be crushed. If white grapes are picked on a hot day, they should be chilled to less than 20°C prior to crushing, but it may be preferable to pick them by night. However, this is not the same for red wine grapes as the fermentation temperature is higher. Fruit destined for white table wine is picked when its sugar content is 23–26° Brix. Grapes for red table wine have a longer hang time. These values are selected such that there is an optimal balance between alcohol yield, flavour and resistance to spoilage. The pH values in these grapes will be 3.2–3.4 and 3.3–3.5, respectively.

Harvesting is increasingly mechanical. While more physical damage occurs, it can be performed under cooler night-time conditions which is desirable, especially for white cultivars. Sulphur dioxide may be added during mechanical harvesting.

Payment is made on the basis of the measured Brix content of the fruit, measured by a hydrometer or, more usually, by a refractometer. A commercial specification will also state the maximum weight of non-grape material that can be tolerated (perhaps 1–2%) and that the berries should be free from mould and rot. For many winemakers, it has been decided that growing their own grapes is prudent. However, the buying in of some material from other suppliers does allow financial flexibility.

The structure of the grape is illustrated in Fig. 3.5. The main features are the skin and the flesh. The skin comprises an outer 1-cell deep epidermis and an inner 4–20-cell deep hypodermis, which is the origin of the colour and most of the flavour compounds in the grape. Sugar and acid are concentrated in the flesh. The sugar content may reach as high as 28%. Tartaric and malic acids account for 70% of the total acids in the grape.

## Grape processing

Nowadays the vessels used for extracting grapes and fermenting wine are fabricated from stainless steel and are jacketed to allow temperature regulation.

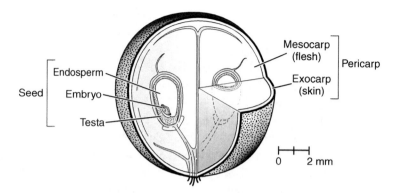

**Fig. 3.5**  The structure of the grape.

These tanks are subject to in-place cleaning, usually a caustic regime incorporating sequestering agents, followed by the use of sanitisers.

Grapes are moved by screw conveyors from the receiving 'bin' to the stemmer-crusher. They pass from there either to a drainer, a holding tank or (in the case of red grapes) directly to the fermenter.

### *Stemming and crushing*

Stems are not usually left in contact with crushed grapes so as to avoid off-flavours. This is not uniformly the case. Pinot noir, then, is some-times fermented in the presence of stems in order to garner its distinct peppery character.

Stemmer-crushers frequently employ a system of rapidly spinning blades, but may have a roller-type design (Fig. 3.6). In either case, there is an initial crushing into a perforated drum arrangement that separates grape from stem.

The aim is even breakage of grapes. If grapes are soft or shriveled, they are tougher to break open. Excessive force will lead to too much skin and cell breakage, and in turn in the release of unwanted enzymes and buffering materials that maintain too high a pH. There will also be problems later on in the clarification stage. It is also important to avoid damaging seeds in order that tannins are not excessively released.

It is not necessary to separate the juice from skins immediately for red wine, but is so for white or blush wines. The colour is located in the skin as polyphenolic molecules called anthocyanins. Blush wines are lighter than rose wines. For the latter, overnight contact between juice and skin with a modest fermentation (perhaps a fall in Brix of 1–5) allows the appropriate extraction of anthocyanins. After rose or blush juice has been separated from the skins, it should be protected from oxidation by the addition of sulphur dioxide ($SO_2$). $SO_2$ addition to the crusher depends on several factors, notably whether mould or rot is present and also what the surface area to volume ratio is in the tank (i.e. the likelihood of air ingress). If the grapes are not infected

**Fig. 3.6**   (a) Grape receiving area, Livingston Winery, California; (b) destemmer and (c) crush pit receiving grapes from gondolas. All photographs courtesy of E & J Gallo.

and the area to volume is low, then $SO_2$ may perhaps be avoided. However, in this instance, the juice should be settled at a low temperature ($<12°C$). The rapid separation of skin and juice for white wines also minimises the pick-up of astringent tannins. The process may also impact other flavour compounds, for example, the flavours that impact Muscat. For certain grapes/wines, therefore, there is a balance to be maintained in terms of oxygen availability, $SO_2$ use, contact time and temperature.

Although seldom used for wines of quality, 'thermovinification' may be used to enhance colour recovery in some wines. The technique involves rapid heating and cooling of crushed grapes. The heat kills the cells, allowing pigments to be released, which may result in undesirable flavours.

Botrytis (see earlier) produces an enzyme called laccase that oxidises red pigments, developing a brown colouration (see Enzymatic browning in Chapter 1). In these circumstances, heating before vinification may be used to destroy the enzyme. Another enzyme that oxidises polyphenols – PPO – is located in the grape *per se*, but it is inhibited by $SO_2$.

During fermentation, the pH should be maintained below 3.8. Wines then tend to ferment more evenly, there is a reduced likelihood of malolactic fermentation and the wine develops better sensory properties. Furthermore, at higher pH, $SO_2$ is less inhibitory to wild yeast. Maintaining this low pH is especially important for white wines. Prolonged contact with the grape skin causes lower total acidity through precipitation of potassium acid tartrate. The pH may be lowered to 3.25–3.35 by the addition of tartaric acid.

### Drainers and presses

Drainers are basically screen-based systems (Fig. 3.7). Presses differ according to the severity of their operations (Fig. 3.8). Membrane or bag presses are very

**Fig. 3.7**  Inside a Diemme Millenium 430 Bladder press, showing drain channels. Courtesy of E & J Gallo.

**Fig. 3.8** Diemme Millenium 430 Press. Courtesy of E & J Gallo.

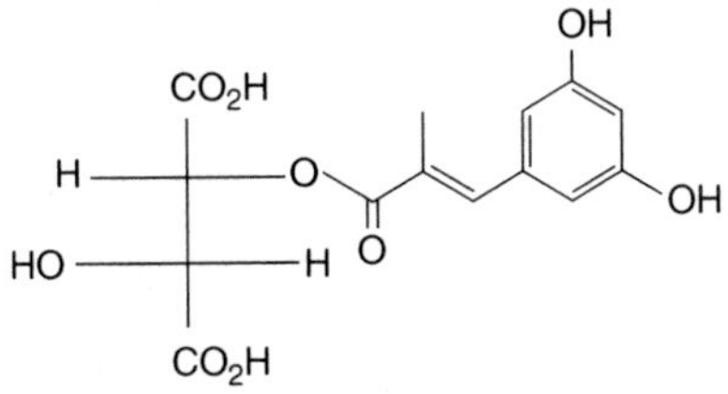

**Fig. 3.9** Caftaric acid.

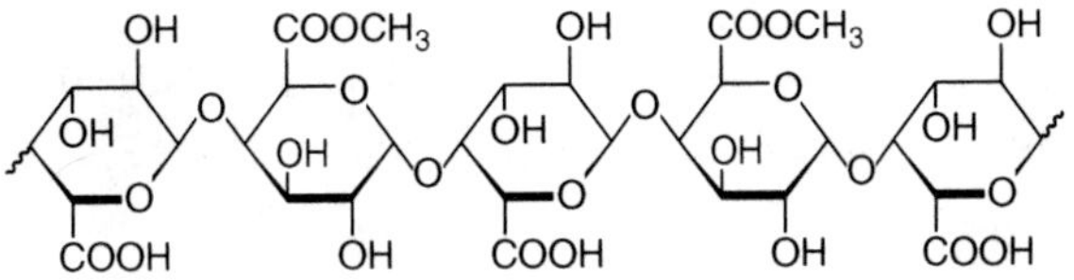

**Fig. 3.10** The repeating unit of pectin: lengthy sequences of anhydrogalacturonic acid partly esterified with methanol.

gentle and leave little sediment. By contrast, bladder presses are often used on account of their rapidity, but the juice tends to contain higher solids levels.

The extent to which Maillard reactions can occur during processing is controlled by attention to temperature, pH and the type of sugar. These reactions occur for the most part at around 15% moisture.

Oxidative reactions may occur, with the major substrates being caffeoyl tartaric acid (caftaric acid; Fig. 3.9), *p*-coumaroyl tartaric acid and feruloyl tartaric acid. These are the precursors in PPO-catalysed browning reactions for those wines that have minimum skin contact.

To accelerate juice settling so as to obtain a clearer product, pectic enzyme is frequently added at the crushing stage to minimise the level of pectin, which originates in the wall material of the grape (Fig. 3.10). The enzyme also allows easier pressing and affords higher yields.

## Fermentation

### *Juice*

Once the juice has separated from the skins, it is held overnight in a closed container. Thereafter it is racked off (or centrifuged), prior to the addition of yeast. Winemakers generally aim to leave some solids as a surface for the yeast to populate (or perhaps as a nucleation site to allow $CO_2$ release, as is the case for the residual cold break in brewery fermentations, see Chapter 2). Failing this, they may add diatomaceous earth or bentonite.

In locations where the grapes do not ripen well owing to a short growing season, it may be necessary to add sugar (sucrose), but only up to a maximum of 23.5°Brix. Such a practice is illegal in some locales, for example, California.

The typical composition of the grapes from which the juice is derived is given in Table 3.3.

Diverse sugars, notably glucose and fructose, are present in essentially equal quantities in mature grapes. Sucrose is hydrolysed at the low pH values involved and this is further promoted by invertase. Total reducing sugars will usually amount to $<250\,g\,L^{-1}$.

The organic acids are predominately tartaric acid in grapes grown in warmer climates and malic acid in grapes from colder climates (Fig. 3.11).

Amino acids and ammonia are present, together with lesser amounts of proteins ($<20$ to $>100\,mg\,L^{-1}$ in the juice). The latter presents a risk to the colloidal stability of wine.

Although vitamins are present in only small amounts, they are generally sufficient for yeast.

A diversity of phenolic compounds is present, and these can be classified as catechins, flavonols and flavanones (Fig. 3.12).

**Table 3.3**   Composition of grapes (percentage of the fresh weight).

| Component | Range |
|---|---|
| Water | 70–85 |
| Glucose | 8–13 |
| Fructose | 7–12 |
| Pentoses | 0.01–0.05 |
| Pectins | 0.01–0.1 |
| Tartaric acid | 0.2–1.0 |
| Malic acid | 0.1–0.8 |
| Citric acid | 0.01–0.05 |
| Acetic acid | 0.0–0.02 |
| Anthocyanins | 0.0–0.05 |
| Tannins | 0.01–0.1 |
| Amino acids | 0.01–0.08 |
| Ammonia | 0.001–0.012 |
| Minerals | 0.3–0.5 |

Information from Amerine *et al.* (1980)

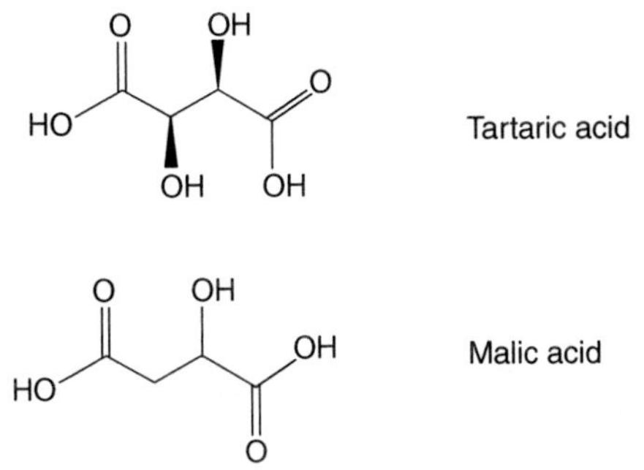

**Fig. 3.11** Grape acids.

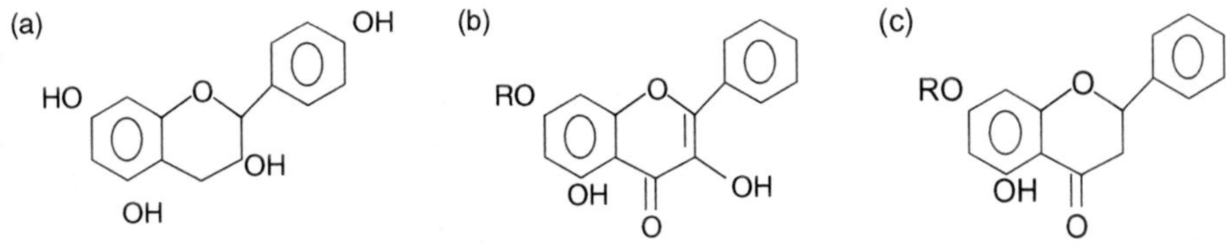

**Fig. 3.12** Some polyphenolic species: (a) catechin, (b) flavonol and (c) flavanone.

**Table 3.4** Yeasts for fermenting wine.

| |
| --- |
| *Saccharomyces cerevisiae* |
| *Saccharomyces bayanus* |
| *Zygosaccharomyces bailii* |
| *Schizosaccharomyces pombe* |
| *Torulaspora delbrueckii* (flor yeast) |

The main inorganic cation in juice is potassium, from $<400$ to $>2000\,\mathrm{mg\,L^{-1}}$.

## Yeast

The relevant species are *Saccharomyces cerevisiae* and *Saccharomyces bayanus* (Table 3.4).

Contrary to commercial-scale brewing, dried yeast is extensively employed in wine making, where the precise nuances of yeast strain seem to be deemed less important than is the case for beer.

Pesticides employed on the grapes can inhibit yeast. Clarification of the must eliminates most of them, but bentonite or carbon treatment may also be employed. However, ironically, the most common inhibitor of fermentation is $SO_2$.

The chief limiting factor in wine fermentations is nitrogen, that is, the amino acid level in the must. Accordingly, it is frequently the case that the level of assimilable nitrogen is increased by the addition of diammonium phosphate.

As for the fermentation of brewer's wort, $O_2$ is introduced to satisfy the demands of the yeast. However, for wine fermentations, aeration is customarily *after* the introduction of yeast so as to avoid the scavenging of the oxygen by PPO.

White wines are fermented at 10–15°C whereas reds are produced at 20–30°C. Fermentation is inherently more rapid at higher temperatures, with the attendant increase in production of flavour-active volatiles such as esters. Rose and blush wines are fermented akin to white wines.

Fermentation tends to be progressively inhibited as the ethanol concentration rises, especially at higher temperatures. Naturally there is also more evaporative loss of alcohol at higher temperatures.

The varietal character of certain wines is better preserved at lower fermentation temperatures. Thus, for example, the terpenols in White Riesling are retained better. As in the case of beer, high levels of the undesirables such as hydrogen sulphide can arise if fermentations are sluggish.

In all cases, fermentation should be complete within 20–30 days. The progress of fermentation is monitored by measuring the decline in Brix value.

Wine is usually racked off the yeast once fermentation is complete. However, some winemakers leave the wine in contact with the yeast for several months, perhaps with intermittent rousing, in order that materials should be released from yeast, beneficially impacting flavour.

Colour and flavour extraction from red grapes is maximised by mixing – either by pumping or by stirring. Usually pumping over (of half of the total vessel contents) is performed twice per day. Extraction is also greater at higher temperatures and increased ethanol concentrations.

A technique traditional for Beaujolais wines is *Maceration carbonique*, which leads to wines with distinct estery, 'pear drop' characteristics. Whole grape clusters are exposed to an atmosphere of $CO_2$. The sugar converts to ethanol (about 2.5% ABV), with the accompanying production of several phenolic compounds. The initial phase of fermentation in the whole grapes is conducted at 30–32°C. The weight of the berries, together with the action of the developed ethanol and carbon dioxide, break down the grape cells and colour is extracted. After 8–11 days, the grapes are pressed and the juice obtained is combined with that which is free running. The whole is fermented to dryness at 18–20°C. Then $SO_2$ is added and the wine is clarified.

## Clarification

White wines are either centrifuged or treated with bentonite, which will also adsorb protein. Bentonite is a clay that contains high levels of aluminium and silica. Sometimes it is substituted by silica gels of the type extensively used in brewing.

Casein may be added to remove phenols, which can also be achieved by PVPP. Isinglass is also sometimes used as a fining agent.

Red wines are primarily fined in order to reduce their astringency. Fining agents include gelatin, egg white and isinglass.

## Filtration

Contrary to most beers, this is relatively uncommon and only performed on an as-needs basis, either to recover wine from lees (i.e. the residual solid material) after cold stabilisation treatments or immediately before bottling. Microbial threats may be eliminated by membrane filtration.

## Stabilization

One of the biggest threats to wine is oxidative browning (see Chapter 1). The ingress of oxygen after fermentation should be minimised. Sometimes 'pinking' of white wines in bottle is prevented by adding ascorbic acid. But the chief antioxidant is $SO_2$, by reacting with the active peroxides in wine

$$H_2O_2 + SO_2 = H_2O + SO_3$$

Metal ions, such as iron, which potentiate the conversion of oxygen into activated forms such as peroxide (see Chapter 1), are removed by casein or citrate.

The sulphur dioxide must be in a free, unbound form at concentrations between 15 and $25\,mg\,L^{-1}$.

Any hydrogen sulphide present in wine may be eliminated by the addition of low levels of copper

$$CuSO_4 + H_2S \rightarrow CuS \downarrow + H_2SO_4$$

Certain inorganic precipitates can be thrown in wine, with tartrate being a key problem. This is avoided by cold treatment of the wine. Protein hazes are avoided by the use of chilling and bentonite.

Maintaining wine in an anaerobic state and with $20$–$30\ mg\,L^{-1}$ $SO_2$ is generally sufficient to prevent spoilage by most bacteria and yeast. Furthermore, when fermented to dryness, most white wines are relatively resistant to spoilage.

## The use of other micro organisms in wine production

Red wines usually undergo a malolactic fermentation, effected by the lactic acid bacteria Pediococcus (homofermentative), Leuconostoc (heterofermentative), Oenococcus (heterofermentative) and Lactobacillus (either). In this process, malic acid is degraded to lactic acid with an attendant decrease in total acidity and a net increase in pH. The bacteria concerned prefer a relatively high pH and tend to be inhibited by $SO_2$. They also do not perform well at too low a temperature. For an effective malolactic fermentation, the wine should have a pH of 3.25–3.5, a total $SO_2$ level below 30 ppm and zero free $SO_2$. The malolactic fermentation formerly depended on the microflora native to the process, but in most instances nowadays the specific bacterial strains required are seeded into the vessel.

Grapes from warm climates tend to contain less malic acid and therefore benefit less from such a fermentation than do grapes from relatively cold areas.

A further type of natural fermentation effected in the production of some wines is the application of certain yeasts (formerly believed to be *Torulaspora delbrueckii* but likelier to be *S. cerevisiae*) growing as a film on the surface in the production of 'flor' sherry. The main impact is the production of acetaldehyde.

## Champagne/sparkling wine

The best such wines are produced from the juice of Pinot noir or Chardonnay grapes. There must be rigorous avoidance of colour development, hence the extensive use of $SO_2$, bentonite and PVPP.

Fermentation in bottle is effected by a culture of *S. bayanus* that is flocculent and able to perform at high alcohol concentrations. The parent wine, invert sugar and yeast are delivered into pressure-resistant bottles sporting a lip for the application of a crown cork. A 2.5-cm headspace will be left in the bottle before it is laid on its side and held at 12°C. The wine will ferment to dryness over a period of several weeks but may be left for more than a year for the achievement of best quality.

There follows the process of 'riddling' in which the yeast is worked into the neck of the bottle. The yeast is loosened by hitting the bottom of the bottle with a rubber mallet or by using a shaking device. Then the bottle is put neck down into a rack at an angle of 45°. The bottle is rotated a quarter turn daily until the yeast sediment has all arrived at the cap. Then the inverted bottle is chilled to 0°C and carried through a brine bath cold enough to yield a frozen plug of wine about 3.5 cm long. The cap is removed and as the ice plug is forced out, it scrapes the yeast with it. The bottle is immediately turned upright again, refilled with wine containing sugar and some $SO_2$, corked and labelled.

In an alternative approach, very cold riddled wine is completely removed from bottles, pooled and cold stabilised under pressure. It is filtered and returned to bottles for corking and labelling as 'sparking wine'.

Certain wines are carbonated simply by bubbling with carbon dioxide prior to packaging (cf. beer).

## Ageing

Contrary to most beers, wines tend to benefit from ageing, which is performed either in tank, barrel or bottle. The extent of ageing is likely to be less for white

**Table 3.5**   Examples of compounds developing in alcoholic beverages aged in oak.

Cyclotene
Dihydromaltol
Ellagic acid
4-Ethylguaiacol
Ethyl maltol
4-Ethylphenol
Eugenol
Furaneol
Furfural
Gallic acid
Hydroxymethyl furfural
$\beta$-Ionone
Maltol
5-Methylfurfural
$\beta$-Methyl-$\gamma$-octalactone
Norisoprenoids
Syringaldehyde
Vanillin

Flavour changes occurring during ageing are not solely due to extraction of substances from the wood. Other significant events include oxidation, evaporation and chemical reactions leading to the production of new compounds.

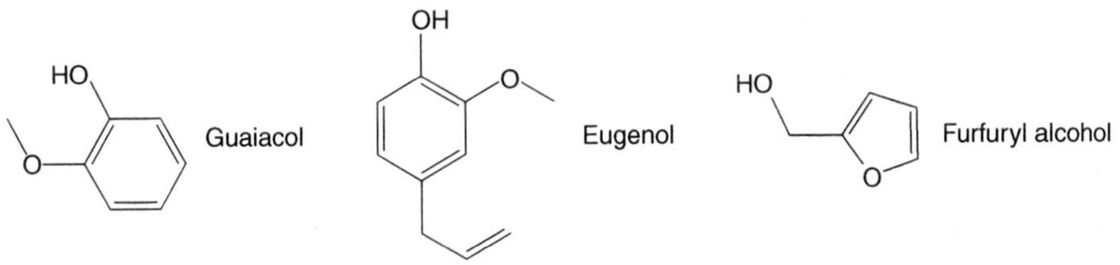

**Fig. 3.13**   Wood-derived flavour compounds.

wines than for reds. During the ageing of wines, there is careful monitoring of colour, aroma, taste and the level of $SO_2$.

The flavour of white wine is very largely determined by the esters produced during fermentation. Some chardonnays are aged in oak barrels, from which some characteristics are derived (Table 3.5). Diverse oaks may be used in ageing, with relevant compounds increasing in level being guaiacol, eugenol and furfuryl alcohol (Fig. 3.13). Burgundy and Loire whites are left on the lees for up to 2 years ('sur lies').

Red wines, having undergone their malolactic fermentation are then aged. Bordeaux wines are held 2 years in barrel. By comparison, zinfandel ageing should not be excessively prolonged in order to retain the raspberry character.

## Packaging

Residual oxygen in wine is removed by sparging with nitrogen gas. Careful control of oxygen levels is effected during the bottling operation *per se*. Some

**Fig. 3.14**   Wine taints.

**Table 3.6**   The major components of table wine.

| Component | Range (g L$^{-1}$) |
| --- | --- |
| Ethanol | 80–110 |
| Methanol | 0–0.3 |
| Propanol | 0.007–0.07 |
| Isobutyl alcohol | 0.007–0.17 |
| Active amyl alcohol | 0.019–0.1 |
| Isoamyl alcohol | 0.08–0.35 |
| 1-Hexanol | 0.001–0.012 |
| 2-Phenylethanol | 0.005–0.07 |
| 2,3-Butanediols | 0.015–1.6 |
| Sorbitol | 0.005–0.39 |
| Mannitol | 0.08–1.4 |
| Erythritol | 0.03–0.27 |
| Arabitol | 0.013–0.33 |
| Glycerol | 1.1–23 |
| Malic acid | 0–6.0 |
| Tartaric acid | 0.5–4.0 |
| Succinic acid | 0.5–1.3 |
| Citric acid | 0–0.3 |
| Acetaldehyde | 0.003–0.49 |
| Acetoin | 0.0007–0.138 |
| Diacetyl | 0.0001–0.0075 |
| Ethyl acetate | 0.001–0.23 |
| Isoamyl acetate | 0–0.009 |
| Mono-caffeoyl tartrate | 0.07–0.23 |
| Mono-*p*-coumaroyl tartrate | 0.008–0.03 |
| Mono-feruloyl tartrate | 0.001–0.016 |
| Various other esters | Various, but low |
| Total amino acids | 0.37–4.2 |
| Protein | 2–2.5 |
| Tannins | 0.05–2.5 |
| Histamine | 0–0.49 |
| Tyramine | 0–0.012 |
| Potassium | 0.09–2 |
| Sodium | 0.003–0.3 |
| Nitrate | 0–0.05 |

Data from various sources.

winemakers add sorbic acid as an antimicrobial preservative for sweet table wines. If such an additive is to be avoided, then more attention must be paid to cold filling and sterility.

## Taints and gushing

Cork taints on wine can come from several sources. Trichloroanisole affords a musty or mouldy character, geosmin an earthy note and 2-methylisoborneol a chlorophenolic aroma (Fig. 3.14). They are due to chlorine treatment of corks with subsequent methylation by bacteria and moulds. It is advisable to keep corks at very low moisture content (5–7%) in order to minimise this problem. Of course metal- or plastic-lined caps do not present this risk – but are widely unfavoured in view of their lesser aesthetic appeal. Taints may also arise from wooden vessels employed in the winery.

Gushing in wine may arise due to microscopic mould growth.

As for beer, the shelf life of wine is greatly enhanced by cool temperature of storage.

## The composition of wine

Table 3.6 presents an approximate summary of the main chemical components of wine.

## Bibliography

Amerine, M.A. & Roessler, E.B. (1983) *Wines: Their Sensory Evaluation.* San Francisco: WH Freeman.

Amerine, M.A. & Singleton, V.L. (1977). *Wine: An Introduction*, 2nd edn. Berkeley: University of California.

Amerine, M.A., Berg, H.W., Kunkee, R.E., Ough, C.S., Singleton, V.L. & Webb, A.D. (1980) *The Technology of Wine Making*, 4th edn. Westport, CT: AVI.

Boulton, R.B., Singleton, V.L., Bisson, L.F. & Kunkee, R.E. (1996) *The Principles and Practices of Winemaking*. New York: Aspen.

Dutruc-Rosset, G. (2000) The state of vitiviniculture in the world and the statistical information in 1998. *Bulletin de l'Office International de la Vigne et du Vin*, **73**, 1–94.

Fleet, H., ed. (1993) *Wine Microbiology and Biotechnology*. Chur: Harwood.

Jackson, R.S. (2000) *Wine Science: Principles, Practice, Perception*, 2nd edn. San Diego: Academic Press.

Waterhouse, A.L. & Ebeler, S., eds (1998) *Chemistry of Wine Flavor*. ACS Symposium Series No. 714. Washington, DC: American Chemical Society.

# Chapter 4
# Fortified Wines

Fortified wines are those in which fermented, partially fermented or unfermented grape must is enriched with wine-derived spirit. According to the European Union (EU) regulations, such liquor wines are those with an acquired alcohol content of 15–22% by volume and a total alcohol content of at least 17.5% by volume.

The chief fortified wines are sherry (originating in Spain, notably Jerez de la Frontera, which is in the southern province of Cadiz), port (from Portugal and made from grapes produced in or around the upper valley of the River Douro in the north of the country) and madeira (from the Portuguese archipelago of Madeira).

The wine fortification technology originated in such regions because the local soil and climate were not well suited to the production of wines of inherent excellence. The process also allowed protection against microbial infection during the storage and shipment of products.

Sherry is only made from white grapes, but port and madeira may be produced from either red or white grapes. In no instance is a single product made from a mixture of the two grape types. Wines upon which sherry is based tend to be dry and the fortification occurs post-fermentation. If the sweetness needs to be increased, it is through the addition of grape-derived products downstream. Such additions usually comprise wines that have been fortified at the start of fermentation: by adding alcohol at the start of fermentation, yeast action is arrested (see discussion in Chapter 2 on yeast stress) and accordingly there is retention of sugar.

Port is usually fortified approximately halfway through the primary fermentation, and so tends to be sweeter than sherry through the preservation of unfermented sugars.

Madeira may be fortified through either route depending on the sweetness targeted in the product.

The wines used to make sherry derive much of their character from ageing in oat 'butts'. Sometimes, however, there is the development of *flor*, a film of yeast on the surface. This yeast may comprise the primary fermenting yeast but may also include other adventitious yeasts from diverse genera.

In contrast, characteristics derived from the grape are substantially more important for wines going into port, especially red port. Much of the character of madeiras develops in the *estufagem* process, which is a heating of the product at, say, 50°C for 3 months.

Sherry, port and madeira are each blended to the target quality during maturation.

Sherry and madeira are fortified using an essentially neutral spirit containing at least 96% ABV and which is continuously distilled from the wine or from related products (the lees or the pomace). Fortification of port is with wine spirit (76–78% ABV). This spirit does contain substances such as alcohols, esters and carbonyl-containing compounds that contribute directly to the flavour of port.

## Sherry

The reader is referred to Reader and Dominguez (2003) for comprehensive details of grapes and vinification techniques; however, these are only subtly different from those employed generally for wines (see Chapter 3).

Nowadays fermentation is likely to be in open cylindrical tanks (500–1000 h L) regulated to ca. 25°C. Rather than employing pure cultures of yeast, starters are prepared using the natural flora on a proportion of grapes harvested before the vintage, the harvest being complete towards the end of September. The initial population will include Hanseniaspora but *S. cerevisiae* soon predominates. Fermentation is completed to dryness by November and a malolactic fermentation will have been effected by endogenous lactic acid bacteria.

Post-fermentation, the young wines are racked from the lees and fortified with spirit (>95% ABV) produced by the distillation of wine and its by-products (lees, pomace). The spirit is first mixed with an equal volume of wine and settled for some 3 days before using to fortify the main wine. This procedure leads to less generation of turbidity than does addition of undiluted spirit.

The young unaged wines are classified into either *finos* or *olorosos* depending on their characteristics. Finos are dry, light and pale gold in colour and have an alcohol content of 15.5–16.5%. They are matured under flor yeast, which tends to develop when the grapes are exposed to cool westerlies when grown on soils rich in calcium carbonate. Olorosos, which are matured in the absence of flor yeast, are dry, rich dark mahogany wines with full noses and alcoholic contents of 21%. The higher levels of polyphenolics in these wines suppress flor development.

Newly fermented wines are left to mature unblended for approximately 1 year. They then pass to a blending process (the 'solera' system), in which the aim is to introduce product consistency. It comprises a progressive topping up of older butts of wine with younger wines (much in the way that balsamic vinegar is derived – see Chapter 9).

A sherry must be aged for a minimum of 3 years before sale. During ageing, flor prevents air from accessing the sherry, and so microbial spoilage and oxidative browning is prevented. If there is no flor, as in olorosos, then oxidative browning can occur. Amontillado sherries are produced with an initial flor maturation followed by ageing in the absence of flor, so oxidation and esterification reactions are prevalent in that style of sherry. The flor process leads to a decrease in volatile acidity and glycerol, as well as an increase in the level of acetaldehyde, the latter meaning that fino sherries have a distinct apple note. Other flavour compounds associated with sherries include 4,5-dimethyl-3-hydroxy-2-($^5$H)-furanone, which affords a nutty character to sherry matured under flor and *trans*-3-methyl-4-hydroxyoctanoic acid lactone, which emerges from the oak and offers the woody note found in many sherries.

Fino sherries are not usually sweetened, are matured for 3–8 years and have alcohol contents of 15.5–17% ABV. Olorosos and Amontillados are generally sweetened and reach 17–17.5% ABV.

Sherries may be fined, traditionally with egg white although increasingly with isinglass or gelatin. They may be centrifuged before filtering and may also be stabilised by treatment with bentonite.

Finally they are cooled through a heat exchanger and ultra-cooler to reach a temperature between −8°C and −9°C, holding there for 10–14 days to chill out colloidally unstable material. Finely ground potassium bitartrate may be added to promote the nucleation of this material. Finally, the sherry is membrane-filtered to eliminate microbes and some solids, prior to bottling.

## Port

The reader is again referred to Reader and Dominguez (2003) for more details on vineyard processes.

Much of the port produced these days is fermented in closed tanks at ca. 16°C with facility for turning the contents. Must is run-off after 2–3 days of fermentation at which point most of the sugars have been converted into alcohol. Fermentation is inhibited by the addition of grape brandy with wine becoming port officially at 19–20% ABV.

Red wines destined for ruby will have been aged for 3–5 years in wood. Those going to tawny will have been aged in wood for more than 30 years. Vintage ports are from wines of a single harvest that are judged to be of outstanding quality. They will be aged in wood for 2–3 years and then the ageing completed in bottle for at least 10 years.

A major contributor to the ageing changes in ruby and tawny is the polymerisation of anthocyanins. This is not only partly through oxidative cross-linking, but also through that induced by acetaldehyde. Other significant aldehydes include the furfurals and lignin degradation products from wood, such as vanillin, syringaldehyde, cinnamaldehyde and coniferaldehyde (Fig. 4.1). Phenols such as guaiacol, eugenol and 4-vinylphenol are also extracted from wood during maturation. Other changes include increases in the level of glycerol and decreases in the levels of citric acid and tartaric acid,

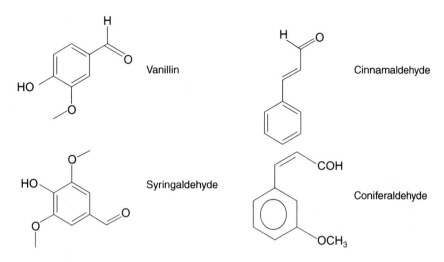

**Fig. 4.1**  Wood-derived species in port.

the latter by the deposition of potassium hydrogen tartrate. In the acidic, high ethanol wines, esters are produced by the reaction of ethanol with acetic, lactic, malic, succinic and tartaric acids.

Ports are blended, especially the ruby's. They are clarified with gelatin, casein or egg white. White ports will be treated with bentonite, and centrifugation is sometimes employed. Rubies and younger tawnies are cold stabilised by holding at −8°C for 1 week. Alternatively, the chilled wine is passed continuously through a crystallising tank containing a concentrated suspension of crystals of potassium bitartrate. Then the wine is filtered with diatomaceous earth followed by sheet-, cartridge- or membrane filtration.

## Madeira

Fermentation may be in various types of vessel, ranging from wooden casks to stainless steel fermenters, but generally there is no temperature control, so 35°C may be reached or perhaps exceeded. Starter cultures are not employed. Fortification to 17–18% ABV is either immediate, to prevent malolactic fermentation and the action of endogenous acetic acid bacteria, or delayed 2–3 months, in which case volatile acidity is likely to have increased.

The heating stage is effected after increasing the sweetness by approximately 2–9°Brix using either a fortified grape juice, concentrated grape must or hydrolysed corn syrup. Heating is by circulating hot water around the product, either using a stainless steel coil in the tank or through a jacket. Heating is typically in concrete at 40–50°C for at least 3 months. A brown hue is produced, together with caramelisation aromas and a soft palate arising from the impact on phenolics. The estufagem process must be conducted during the first 3 years.

Madeiras are mostly aged in wood. Vintage madeiras must come from a single variety in a single year and must be aged for more than 20 years in wood and at least 2 years in bottle. Blending of madeira is a simplified version of the port system.

Many madeiras are charcoal-treated to remove the more extreme characteristics developed during the heating stage. They are fined with casein, treated with bentonite and held at $-8°C$ for 1 week before filtration using diatomaceous earth and ensuing sheet or sheet-plus cartridge filtration.

## Bibliography

Fletcher, W. (1978) *Port: An Introduction to Its History and Delights*. London: Philip Wilson.

Fonseca, A.M., Galhano, A., Pimental, E.S. & Rosas, J.R.-P. (1984) *Port Wine. Notes on Its History, Production and Technology*. Oporto: Instituto do Vinho do Porto.

Gonzalez, G.M. (1972) *Sherry, the Noble Wine*. London: Cassell.

Jeffs, J. (1992) *Sherry*. London: Faber and Faber.

Reader, H.P. & Dominguez, M. (2003) Fortified wines: sherry, port and madeira. *Fermented Beverage Production*, 2nd edn. (eds A.G.H. Lea & J.R. Piggott), pp. 157–193. New York: Kluwer/Plenum.

Robertson, G. (1992) *Port*. London: Faber and Faber.

Suckling, J. (1990) *Vintage Port*. San Francisco: Wine Spectator Press.

# Chapter 5
# **Cider**

Cider is an alcoholic drink produced by fermenting extracts of apples, though in the United States the term generally describes a non-alcoholic product, with the alcoholic version being termed 'hard cider' and produced in such apple-growing states as New England and upstate New York. Much of the latter is actually produced for direct conversion into vinegar.

In this chapter, I focus on cider making in the United Kingdom, but it is important to stress that cider is also important in France (Normandy and Brittany), Germany (the Trier/Frankfurt area) and Northern Spain, each of which has some individual manufacturing approaches.

Perry is the equivalent product made from pears, but production of this is on a far smaller scale. Both of these products have a pedigree stemming back at least to the days of Pliny in the Mediterranean basin. Cider production probably came to England from Normandy even before 1066.

The United Kingdom is the biggest producer of cider. Historically the major production areas have been the West Midlands, notably the counties of Hereford and Worcester, Gloucestershire, Somerset and Devon. Smaller amounts have been produced in East Anglia, Sussex and Kent.

In the earliest days of cider production in England, it achieved such a high status that it was a peer for wines. However, particularly during the nineteenth century, its quality declined and it assumed the status of being a low-cost source of alcohol for peripatetic farm workers. The 'scrumpy' image was assumed. However, in the late twentieth century, cider once more gained appeal as a drink of quality, including for young people.

The biggest selling style of cider is as a clear carbonated, light flavoured beverage in bottle or can with an alcohol content of between 1.2% and 8.5% ABV. Increasingly there is a trend towards chaptalisation – that is, the addition of sugars or syrups prior to fermentation to supplement the carbo-hydrate derived from apple. For the most part, modern ciders may comprise only 30–50% apple juice.

New product development has been rife in the cider industry in recent years. Thus, *inter alia* there have been higher alcohol variants, 'white' ciders stripped of their colour, so-called ice versions (cf. beer) and ciders flavoured with diverse other components.

When served on draught, cider is essentially a competitor for beer, pri-marily the lager-style products. However, there are styles of draught cider that are much more akin to cask conditioned ales. Nonetheless, there is probably a closer match between cider making and wine making than there is with brewing.

In France, ciders tend to be of lower alcohol content and distinctly sharp in flavour. Those from the Asturias region of Spain are somewhat vinegary and foamy, while those from Germany tend to have relatively high alcohol content.

# Apples

The starting material for cider production is raw apples. A classification for these is offered in Table 5.1.

It is not necessarily the case that cider must be made from true cider apples. For example, cider has been made successfully from Bramley apples. Frequently the substrate derived directly from the apple is supplemented with Apple Juice Concentrate (AJC).

There are several advantages to using true cider varieties. They tend to have high sugar contents, of up to 15%. They display a range of acidities, from 0.1% to 1%. Their fibrous structure makes it easier to effect pressing and with higher yields of juice. It is possible to store them over a period of several weeks without losing texture, during which period their starch converts into sugar. Finally, they have a high tannin content (perhaps ten-fold higher than in dessert apples), this being important for body and mouthfeel. The polyphenols also inhibit breakdown of pectin, rendering the pulp from bittersweet apples less slimy and therefore easier to process.

The polyphenolics in apples comprise a range of oligomeric procyanidins based on the flavanoid (−)-epicatechin (Fig. 5.1). Also present are the phenolic acids chlorogenic and *p*-coumaroyl quinic acid, as well as the glycosides, phloretin glucoside and xyloglucoside (Fig. 5.2).

**Table 5.1** Types of cider apples.

| Type of apple | Tannin content (%) | Acid content (%) |
|---|---|---|
| Bittersharp | >0.2 | >0.45 |
| Bittersweet | >0.2 | <0.45 |
| Sharp | <0.2 | >0.45 |
| Sweet | <0.2 | <0.45 |

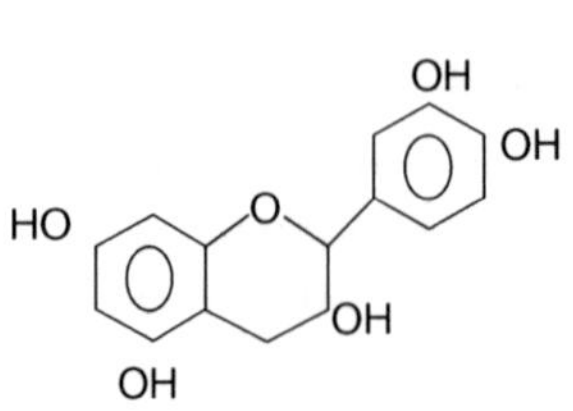

**Fig. 5.1** Epicatechin.

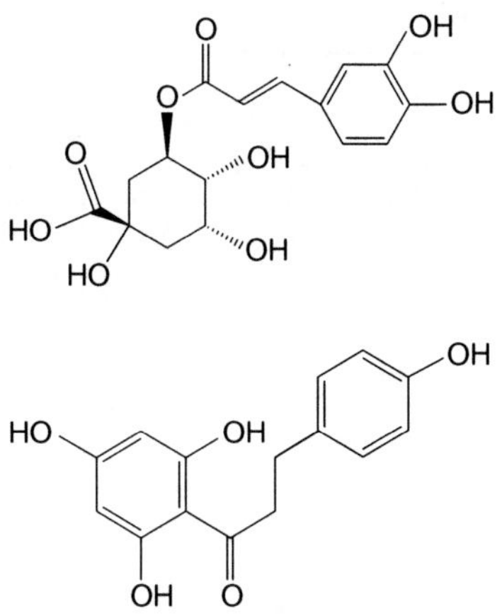

**Fig. 5.2** Phenolic species derived from apples.

The cider orchards are different for cider apples. The aesthetic appeal of the appearance and size of the fruit is relatively unimportant when compared with apples that are intended to be sold as eating fruit. Of more significance is the ease with which they can be harvested. The apples are for the most part grown on bush trees with more than 30 per acre (cf. 20 per acre for dessert apples). Cropping is biennial.

Most of the larger cider making companies possess their own orchards. They also enter into contracts with outside growers for a proportion of their raw material. Cider is usually produced from more than a single cultivar in order to achieve the preferred balance of acidity, sweetness and astringency/bitterness (Table 5.2). The gross composition of cider varieties is actually not very dissimilar to that of other apples and leads to a pressed juice with an overall composition depicted in Table 5.3.

The most likely limiting factor will be the assimilable nitrogen content, depending on the nutrient status of the trees in the orchard. By contrast, the total polyphenol content of apples tends to be inversely related to this nutrient status.

AJC is now extensively used in cider making. Typically it has a concentration of 70° Brix, the high osmotic pressure meaning that it can be stored for long periods and therefore purchased at economically favourable times. Sometimes, however, AJC made from true bittersweets is in short supply and it may be produced in-house. Alternatively, the apple juice may be supplemented with cane or beet sugar or hydrolysed corn syrup.

## Milling and pressing

Apples are used when fully ripe and are customarily stored for several weeks so as to convert all of the starch into fermentable sugar. The apples are sorted and washed with the aim of eliminating debris and any rotten fruit.

**Table 5.2**   Cider apple cultivars.

| | |
|---|---|
| *Bittersharp* | *Sharp* |
| Brown Snout | Brown's Apple |
| Bulmer's Foxwhelp | Frederick |
| Chisel Jersey | Reinette Obry |
| Kingston Black | |
| | |
| *Bittersweet* | *Sweet* |
| Ashton Brown | Northwood |
| Chisel Jersey | Sweet Alford |
| Dabinett | Sweet Coppin |
| Ellis Bitter | |
| Harry Master's Jersey | |
| Major | |
| Medaille d'Or | |
| Michelin | |
| Taylor's | |
| Tremlett's Bitter | |
| Vilberie | |
| Yarlington Mill | |

**Table 5.3**   Major components of cider apple juice.

| Component | Range |
|---|---|
| Fructose | 70–110 g/L |
| Glucose | 15–30 g/L |
| Sucrose | 20–45 g/L |
| Pectin | 1–10 g/L |
| Amino acids | 0.5–2 g/L |
| Potassium | 1.2 g/L |
| pH | 3.3–3.8 |
| Phenolics and polyphenolics | 1–2.5 g/L |

Derived from Lea & Drilleu (2003).

Formerly the apples were crushed by stone or wooden rollers with an ensuing pressing in rack and cloth. The pulp was layered in woven synthetic clothes that alternated with wooden racks, the arrangement being referred to as a 'cheese'. Straw was used to separate the layers. The cheese was then stripped down and the pomace mixed with water 10% by weight before re-pressing. The residual pomace was used as animal feed or for pectin production.

In modern cider making facilities, a high-speed grater mill feeds a hydraulic piston press. Within the press are compressible chambers (cf. the mash filters employed in brewing), with many flexible ducts that are enclosed in nylon socks. When the piston is compressed, it forces juice through the ducts. There may be a second extraction by water. When the piston is withdrawn, the dry pomace falls away readily. Yields are much higher (75%+) and there are much lower levels of suspended solids in the apple juice.

The juice is afforded a coarse screening before it is run to tanks fabricated from fibreglass, stainless steel, polyethylene or wood.

## Fermentation

Some blending of juices may occur prior to fermentation and additions made. In particular, there may be a blending with sugars or AJC, to arrive at a specific gravity of 1.08–1.1. The FAN level may be raised to $100\,mg\,L^{-1}$ by the addition of ammonium sulphate or ammonium phosphate. Thiamine may be added, perhaps at 0.2 ppm, but this must be separate from the addition of sulphite as the latter will destroy it. Other B vitamins that are required are pantothenate (2.5 ppm), pyridoxine (1 ppm) and biotin (7.5 ppb). Such materials are especially likely to be limiting if the cidermaker is using significant quantities of AJC or sugars.

Another potential problem with AJC is the generation of O- and N-containing heterocyclics within it (by Maillard reactions – see Chapter 1), which are inhibitors of yeast. They can be removed by the treatment of AJC with activated charcoal. If the apple juice and its additions are too 'bright', then it will be necessary to add some solids (e.g. bentonite) to act as nucleation sites, the escaping $CO_2$ relieving inhibition of the yeast and also serving to maintain agitation in the fermenter. We have already encountered this for the fermentations of beer and wine.

Pectolytic enzymes are sometimes added to initial fruit pulp or to the juice immediately prior to fermentation.

$SO_2$ is traditionally added to prevent the growth of contaminating microorganisms (Table 5.4). It is less critical from that aspect with the advent of dried wine yeast, but it is still important from a flavour perspective and is not without significance for antimicrobial protection. The effectiveness of $SO_2$ increases as the pH decreases because it is the undissociated form of bisulphite which has the antimicrobial properties. The pH is lowered to less than 3.8 by the addition of malic acid prior to the addition of sulphite.

Healthy fruit generally will only contain low levels of sulphite-binding agents and should have sufficient $SO_2$ to offer effective resistance to spoilage before addition of yeast. If, however, the fruit is in less good condition, then it

Table 5.4  The quantity of sulphur dioxide that should be added to cider apple juice.

| pH | $SO_2$ to be added ($mg\,L^{-1}$) |
|---|---|
| 3.0–3.3 | 75 |
| 3.3–3.5 | 100 |
| 3.5–3.8 | 150 |
| >3.8 | 150 (after blending or acid addition to achieve a pH < 3.8) |

Based on Lea & Drilleu (2003).

may contain materials such as 5-ketofructose or diketogluconic acid as a result of bacterial activity. This type of substance binds $SO_2$ and therefore reduces the endogenous protectant level. Furthermore, if ascorbic acid is oxidised to 1-xylosone, this also binds $SO_2$. Finally, if AJC is depectinised, this will yield galacturonic acid that will also diminish $SO_2$.

In traditional cider making, the yeast was delivered adventitiously with the fruit or the equipment (Saccharomyces does not naturally inhabit cider apples – but it is to be found on presses). $SO_2$ suppresses the growth of most microbes other than Saccharomyces. Traditionally a succession of microflora in juice that had not been sulphited was involved in metabolising apple juice to cider. Saccharomyces was significant relatively late in the process. The introduction of $SO_2$, however, rendered Saccharomyces as being vastly more important in the process. Since the 1960s, though, the vast majority of cider fermentations have been seeded. Juice should be held at $<10°C$ prior to the addition of that yeast in order to prevent native flora from kicking off fermentation. Many of the cultures now added were originally isolated from the cider factories themselves, but some cidermakers use wine yeasts with well-defined characteristics, including the spectrum of flavour compounds that they produce and their flocculation behaviour. Since the 1980s, there has been widespread use of active dried wine yeast, which simply needs mixing with warm water, freeing the cidermaker from the need for in-house propagation. Some will employ an aerobic yeast incubation period so as to ensure that the yeast membranes are in good condition in order that the yeast will be capable of effecting very high levels of alcohol production.

Frequently the inoculum is a mixture of *Saccharomyces pastorianus* and *Saccharomyces bayanus*. The former is felt to give a lively start to the fermentation, whereas the latter performs better later in the process, and ferments to dryness.

Where temperature control is effected (this is not universal), this is likely to be within the range 15–25°C. If the fermentation displays sluggishness, then a portion of the goods may be warmed to 25°C by pumping through a heat exchanger. Most fermentations will be complete in 2 weeks.

Ciders are subjected to a malolactic fermentation as in the case of some wines (see Chapter 3). This is effected by heterofermentative *Leuconostoc oenos*, together with other lactobacilli. This is favoured if there is no sulphiting in fermentation and storage and also by autolysis of yeast when the cider is allowed to stand unracked on its lees. As sulphiting is so widespread these days, the malolactic fermentation is probably less significant than it once was. Furthermore, there is a lessening tendency to leave cider on the lees. In the malolactic fermentation, there is a conversion of malic to lactic acid and the release of carbon dioxide. The resultant cider will tend to have a more rounded, complex flavour that is less acidic. The process is inhibited if the pH is too low.

A range of sulphite-binding compounds are produced during fermentation, but the most potent binder of $SO_2$ is acetaldehyde (Fig. 5.3). Essentially,

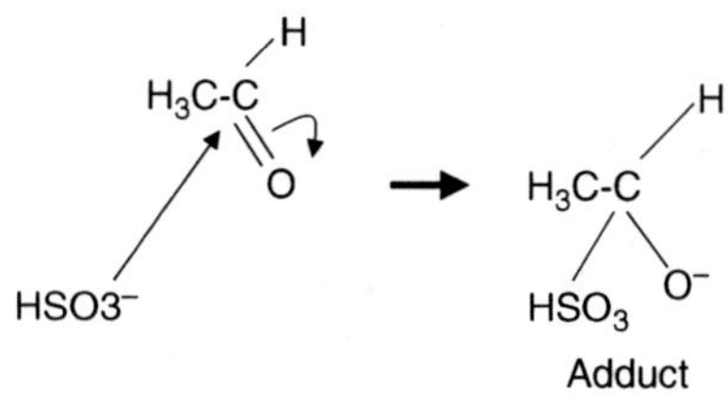

**Fig. 5.3** Adduct formation.

until all of this is bound to $SO_2$, no free $SO_2$ can remain to bind other components. Indeed, $SO_2$ bound to carbonyls such as acetaldehyde has little antimicrobial action, which is why cidermakers try to minimise the level of carbonyls. The addition of thiamine reduces the production of pyruvate and of $\alpha$-ketobutyrate. Pantothenate can reduce the production of acetaldehyde.

## Cider colour and flavour

The colour of cider arises through the oxidation of polyphenols in the juice. It can be regulated by the addition of sulphite. If the latter is added immediately after pressing, then nearly all colour development is suppressed due to binding of sulphite to the quinoidal forms of the polyphenolics. If $SO_2$ is added later, there is less reduction of colour because the quinones have become more intimately cross-linked. The colour decreases during fermentation because of the reducing nature of yeast.

Maillard browning reactions can occur during the storage of AJC, and these coloured products cannot be dealt with by yeast.

The colour of finished cider is standardised by the addition of caramel or other permitted colorants. The colour is removed from speciality products like white ciders by the action of adsorbents such as activated carbon.

The traditional high bitterness and astringency of ciders originate with the procyanidins. Procyanidins with a degree of polymerisation (DP) 2–4 are bitter and are referred to as 'hard tannins'. Those with a DP of 5–7 are astringent ('soft tannins'). The relative delivery of bitterness and astringency depends both on the apple cultivar and on how the apples are processed. Oxidised polyphenols adsorb (become tanned) onto the apple pulp and this suppresses both astringency and bitterness. If oxidation occurs in the absence of the pulp, then there is a relative transition from bitterness to astringency as the units polymerise. Alcohol tends to enhance bitterness but suppresses astringency.

As in the case of beer and wine, the yeast produces a range of volatile components (e.g. esters), and key variables are yeast strain, fermentation temperature, and the clarity and nutrient composition of the fermentation feedstock. Higher quality apple cultivars tend to give juice containing lower levels of assimilable nitrogen, and the attendant slower fermentation rates may be associated with enhanced flavour delivery. For instance, levels of 2-phenylethanol may be increased. Cloudy juices will ferment to give increased levels of fusel oils.

There are several non-volatile glycosidic complexes in apples that are hydrolysed by endogenous glycosidases when the fruit is disrupted. The malolactic fermentation results in the production of diacetyl which can afford a desirable buttery note to some ciders.

Spicy and phenolic notes arise from ethylphenol and ethyl catechol that come from phenolic acid precursors (Fig. 5.4). These are major contributors to the bittersweet flavours of well-made traditional ciders. However, at high levels, they give characters reminiscent of barnyards, possibly due to the slow growth of Brettanomyces in storage.

A listing of volatile components present in cider is offered in Table 5.5.

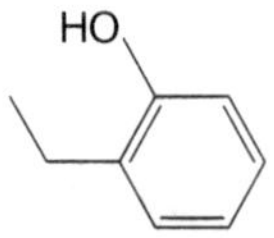

2-Ethyl phenol

**Fig. 5.4**    A source of spiciness in cider.

**Table 5.5**    Volatile constituents of cider.

| | |
|---|---|
| Iso-amyl alcohol | Methionol |
| Benzaldehyde | 2-Methyl-butan-1-ol |
| Iso-butanol | 3-Methyl-butan-1-ol |
| *n*-Butanol | 2-Methylpropanol |
| Decanal | Nonanoic acid |
| δ-Decalactone | Nonanol |
| Decan-2-one | Octanoic acid |
| Diethyl succinate | Octanol |
| Ethyl acetate | Iso-pentanol |
| Ethyl benzoate | 2-Phenylethanol |
| Ethyl decanoate | 2-Phenylethyl acetate |
| Ethyl dodecanoate | *n*-Propanol |
| Ethyl guaiacol | sec-Pentanol |
| Ethyl hexanoate | Undecanal |
| Ethyl-2-hydroxy-4-methyl pentanoate | |
| Ethyl lactate | |
| Ethyl-2-methylbutyrate | |
| Ethyl octanoate | |
| *n*-Hexanol | |
| Hexanoic acid | |
| Hexyl acetate | |

## Post-fermentation processes

Racking consists of removing the newly fermented cider from its lees. In modern cider making, this may occur relatively soon and in the absence of maturation, prior to blending and packaging. More traditional processing has the cider left on the lees for several weeks, with racking into tanks for months of storage with minimum contact with air. The malolactic fermentation may be encouraged, in which case sulphiting is avoided at this point.

Initial clarification of cider is by natural settling, by fining (bentonite, gelatine, chitosan, isinglass), or by centrifugation. Alternatively, a combination of these may be used.

The ciders will be filtered before packaging and may be blended, aided by expert tasting. If fermentation was to a higher-than-target alcohol content, then the cider will be thinned by the addition of water, and sugar or malic acid may be added, as well as of course carbon dioxide.

Final filtration is by powder, sheet and/or membrane filtration. There is increasing use of cross-flow microfiltration (Fig. 5.5). Most ciders are pasteurised and carbonated en route to final pack.

Typically 50 ppm $SO_2$ will be added to give a free $SO_2$ level of 30 ppm, but the precise figures will depend on the level of endogenous binding compounds present in the cider. If the cider is destined for cans, then $SO_2$ levels must be lower because as little as 25 ppm can cause damage to the lacquer layer and to the production of hydrogen sulphide.

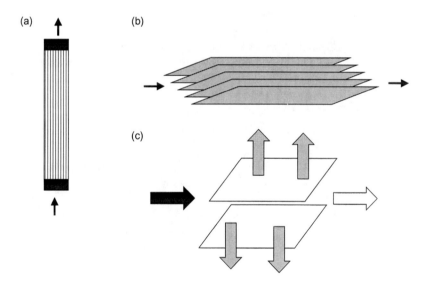

**Fig. 5.5** Cross-flow microfiltration. The cider flows through multiple bundles of porous membranes. Particles, including micro-organisms, are held back by the membranes, with the clarified liquid emerging at right angle to the direction of flow, the continuous nature of which ensures that particles do not adhere to the pores and plug them.

Ascorbic acid may be added, but these days there is less use for sorbic acid as it is only fully effective in the presence of $SO_2$ and, further, it is only active against yeast and not bacteria.

## Problems with cider

Cider sickness, caused by infection through *Zymomonas anaerobia* is now very uncommon, as it is countered by the lower pHs ($<3.5$) and reduced tendency to have residual sugar in the product. Symptoms include an aroma of banana skins and a white turbidity due to the acetaldehyde produced reacting with polyphenols to form insoluble complexes.

Mousiness in cider is due to isomers of 2-acetyl or ethyl tetrahydropyridine (Fig. 5.6) produced by lactic acid bacteria or Brettanomyces under aerobic conditions. Detection of the flavour depends on reaction of the compounds with saliva, with the acidity of the saliva releasing the compounds from the base forms where they are not detected. Thus, simple smelling of cider will not tell whether there is a problem or not.

Ropiness in cider is due to the production by lactic acid bacteria of a polymeric glucan that increases the viscosity of the cider, which appears to be oily when poured due to the movement of the slimy glucan.

Lactic acid bacteria may also break down glycerol. They produce 3-hydroxypropanal which spontaneously dehydrates to generate acrolein that has a bitter taste and a pungent aroma (Fig. 5.7).

Chill hazes in cider are due to complex formation between polyphenols and polysaccharides, and to a lesser extent, with the very low levels of proteins. This is promoted by iron and copper, the levels of which should be minimised.

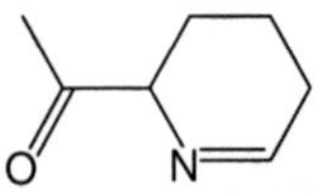

2-Acetyltetrahydropyridine

**Fig. 5.6**   A source of mousiness in cider.

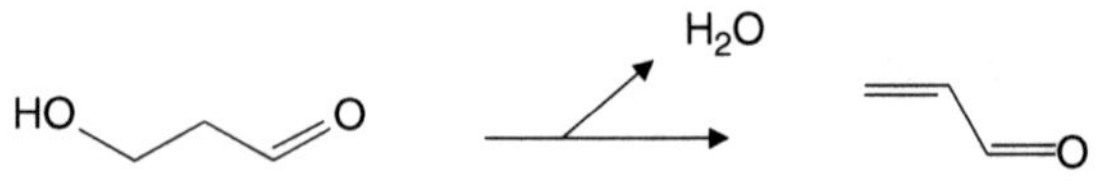

3-Hydroxypropanal                                   Acrolein

**Fig. 5.7**   A source of pungent bitterness in cider.

# Bibliography

Charley, V.L.S. (1949) *The Principles and Practice of Cidermaking*. London: Leonard Hill.

Downing, D.L., ed. (1989) *Processed Apple Products*. New York: AVI Van Nostrand.

Lea, A.G.H. & Drilleu, J.-F. (2003) Cidermaking. In *Fermented Beverage Production*, 2nd edn. (eds A.G.H. Lea, & J.R. Piggott) pp. 59–87. New York: Kluwer/Plenum.

Morgan, J. & Richards, A. (1993) *The Book of Apples*. London: Ebury.

Pollard, A. & Beech, F.W. (1957) *Cider-making*. London: Rupert Hart-Davis.

Williams, R.R., ed. (1991) *Cider and Juice Apples: Growing and Processing*. Bristol: University of Bristol.

# Chapter 6
# Distilled Alcoholic Beverages

The principal distilled beverages are those derived from either grain (whiskies), grapes (cognac, armagnac, brandy) or molasses (rum).

## Whisk(e)y

Whisky (spelled this way for Scotch, but as whiskey for Irish and other forms of the product) is a distilled beverage made from cereals and normally matured in oak. It is subject to a great deal of legislation and custom.

EU regulations state that it can be made from any cereal aided by starch-degrading enzymes with distillation to less than 94.8% ABV, with ensuing maturation in wooden casks of less than 700 L in volume for a period in excess of 3 years for sale at a strength in excess of 40% ABV. UK legislation dictates that Scotch whisky must be produced in Scotland, the enzymes must be entirely derived from malt and the only permitted addition is caramel. The United States, Japan and Canada have their own legislative peculiarities that will not be discussed here.

The major cereals used for the manufacture of whisky are barley, wheat, rye and corn (maize). Malted barley is employed as a source of flavour and enzymes, which are not only responsible for converting the barley starch but also that of adjuncts to fermentable sugars. The main analytical criteria for whisky malts are their diastatic power, $\alpha$-amylase and extract, especially when they are being used alongside adjunct. The malts may be 'peated', that is, flavoured with the smoke from peat burnt on the kiln. Such malts are classified on their content of phenols.

Rye (*Secale montanum*) is quite widely used in Eastern Europe and former USSR, and is sometimes malted. Wheat (*Triticum vulgare*) has largely replaced corn in Scotch grain whiskies as the cost of importing grain from the United States became prohibitive and it is also used in some American whiskies. However, in the United States, corn (*Zea mays*) is especially widely used.

Malt is essentially mashed as in the case for beers, with clear wort being important to prevent burning on the stills. Wort from unmalted grain, however, is not separated from the spent grains because modern continuous distillation processes do not demand it. Fermentation and distillation are effected with all of the grain materials still present.

For malt whisky, mashes of water : grist ratio of 4 : 1 will be mixed in at 64.5°C, the malt having been broken in a roller mill. Although modern malt distilleries are changing over to the use of lauter tun technology (cf. brewing, Chapter 2), traditional distillery mash tuns feature rotating paddles to mix the mash and these will be employed for approximately 20 min before allowing the mash to stand for 1 h. The worts will then be collected before addition of a second water (70°C; 2 m$^3$ per ton) and collection of those worts, followed by waters at 80°C (4 m$^3$ per ton) and 90°C (2 m$^3$ per ton). The first and second worts are cooled by a paraflow heat exchanger to approximately 19°C and diverted to a fermenter or washback. The third and fourth worts are pooled as part of the mashing water for the next mash. Unlike for the brewing of beer, there is no boiling of worts.

The initial processing in the production of grain whiskies is significantly different from that of malt whiskies. Indeed it is not unheard of for distilleries to work with unmilled grain, in which case prolonged cooking is a necessity. For the most part, however, the first stage in production is the hammer milling of the cereal. The desire is fine particles that are readily extracted by water. The cereal is mashed with 2.5 parts water (or recycled weak worts or 'backset', which is a portion of the stillage from the distillation process that has had its solids removed. The latter is felt to deliver yeast nutrients). The mash, typically at 40–45°C, is agitated to ensure that there is no sticking together of grist ('balling'). Some malted barley is likely to be included as a source of enzymes. The slurry is now pumped to a cooker (pressure vessel) wherein the mash is mixed and injected with steam, to achieve gelatinisation of the cereal. The temperature will be raised to 130–150°C and held there for a relatively short period of time. Mixing is essential to avoid charring and excessive browning (Maillard) reactions. The contents of the cooker are now discharged to a flash cooling vessel, the sudden fall in pressure being referred to as 'blow-down'. The impact of this is to release any residual bound starch from the grain matrix. The temperature falls rapidly to around 70°C. The slurry is mixed with a separate slurry of malt (10–15% of the total grist bill) that may be at 40°C, but alternatively may be at the conversion temperature for starch (65–70°C). The malt enzymes then catalyse not only the hydrolysis of the malt starch but also that from the cooked grain. Food grade enzymes will also be added – and to some extent there may still be the use of green (unkilned) malt as a source of enzymes. Mashing will typically be for up to 30 min. Although the wort was formerly separated from the grains, this tends not to be done now in grain distilleries, and the whole mash contents are transferred to the fermenter. There is no boiling, so enzymes can continue to work. Furthermore, it also means that the fermenter contents can be more concentrated than would be the case otherwise . The downside to this is the risk of fouling of stills.

Fermentation of whisky was formerly performed widely with the surplus yeast generated in brewery fermentations. However, specific strains particularly suited to whisky production have been developed and these are supplied by yeast manufacturers in bulk for commercial use. Hybrids emerged not only from the ale strain *Saccharomyces cerevisiae* but also from the 'wild yeast' *Saccharomyces diastaticus*, which produces a spectrum of enzymes fully capable of hydrolysing starch to fermentable sugar. Thus, the distilling strains enable high alcohol yield. The strains may also be selected on the basis of their ability to produce esters.

Yeast is supplied either as compressed moist yeast, as 'cream yeast' (see Chapter 12) or, increasingly, as dried yeast. Quality considerations of the yeast (viability, etc.) are just as for brewing (see Chapter 2).

Fermentation on a small scale may be in closed wooden barrels, but on a larger scale, it will be in stainless steel vessels known as washbacks. Unlike in breweries, there is little temperature control during fermentation, other than to target the initial temperature, which may typically be in the range 19–22°C. The temperature may go as high as 34°C during fermentation, hence the need for ale-based strains rather than lager-based ones. Typically the fermentation is complete within 40–48 h. Some advocate holding a few hours prior to distillation in order to ensure that the endogenous lactic acid bacteria have an opportunity to enhance flavour.

## Distillation

The stills used in the production of whisky are of two types: batch and continuous. Batch (or pot) stills employ double or triple distillation and generate a highly flavoured spirit. Continuous stills provide lighter flavoured spirits that are mostly employed in blending.

Pot stills are traditionally of copper, which may reduce the sulphuriness of the whisky (Fig. 6.1). The still comprises three major parts: the pot, which holds the liquid to be distilled; a swan neck and lyne arm; and a condenser. The precise design of the swan neck/lyne has a considerable impact on the reflux pattern obtained and hence on the flavour.

The pot is heated either directly or indirectly. In the former case, an agitator may be present to prevent charring. Pots can be of diverse shapes, but in traditional Scotch whisky production, there are two stills: the wash still and the spirit still. All of the fermenter contents (the 'wash' will typically be 8% ABV) are transferred to the wash still and boiled for between 5 and 6 h to render a distillate known as 'low wines' which has an alcohol strength of 20–25% ABV. This is subsequently transferred to a smaller spirit still. The spirit coming over from this can be divided into three components: the foreshots, the middle cut and the feints. The charge to the spirit still is a mix of foreshots and feints and low wines to a net alcohol concentration of less than 30% ABV. The foreshots emerge first from the still, the feints last. They contain the undesirable highly volatile and least volatile components, respectively. They are recycled for re-distillation. The foreshots represent perhaps the first 30 min of the distillation and are collected in the feints receiver until the opening distillate strength of 85% has fallen to 75%. At this point, the spirit is judged to be potable and is collected in the spirits receiver. Collection proceeds for up to 3 h, with the alcohol dropping to 60–72% ABV. Thereafter the flow is

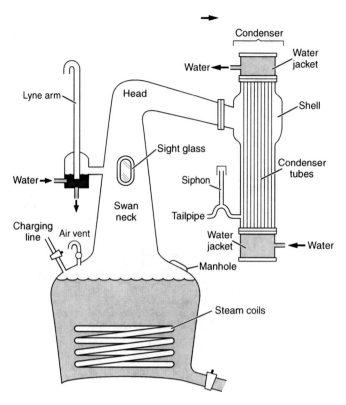

Water

Condenser

Water jacket

Lyne arm

Head

Shell

Sight glass

Condenser tubes

Water

Siphon

Swan neck

Charging line

Air vent

Tailpipe

Water jacket

Water

Manhole

Steam coils

**Fig. 6.1**   A pot still.

diverted once more to the feints receiver and collection may continue until the alcohol reaches as little as 1% ABV.

Continuous distillation takes place in column stills, the most famous of which being that designed by Aeneas Coffey (Fig. 6.2). It comprises two adjacent columns. The wash is preheated by passing it through the tube in the second column (rectifier). Thence it is fed into the first column (analyser) near the top and steam is passed in at the base of the column. As the wash falls, volatiles are stripped from it and these emerge from the top of the column, passing to the rectifier column. Alcohol separates from water at the base. The spirit is removed towards the top of the rectifier. The final cut is taken off from the base of the column. Foreshots (from the top) and feints (from the base) are recycled into the top of the analyser.

Inside the column is a series of plates with holes that permit the upwards flow of vapour. The plates are linked by downcomers that alternate on opposite sides of the plates such that the descending liquid is obliged to flow across each plate. After distillation, new distillates are diluted (e.g. to 58–70% ABV) before filling in oak casks.

The residue from the distillation process is called 'pot ale'. In grain distilleries, it is mixed with spent grains and yeast, whereas in malt distilleries, it is blended with grains and thence despatched for animal feed.

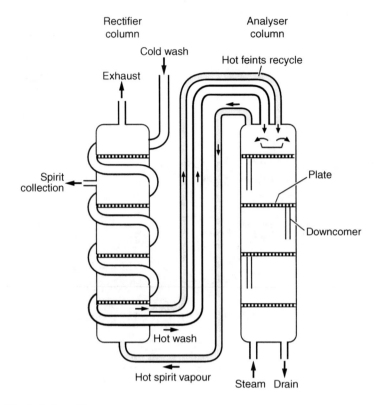

**Fig. 6.2**   A Coffey still.

Whiskies are matured in oak casks. Whereas American bourbon and rye whiskies are put into new oak casks, Scotch, Irish and Canadian whiskies are filled into casks that have previously been employed for Bourbon or for sherry. For the most part they comprise 50 L butts. Whisky casks are either of American white oak (which are used for Fino and Amontillado Sherries) or Spanish Oak (used for Oloroso Sherry). The bourbon casks used for Scotch whiskies must be filled at least once with bourbon and the whiskies must have been in the cask for at least 4 years. Ageing of whisky in most countries must be for at least 3 years. There is a significant loss of alcohol by evaporation in this time, referred to as the 'angel's share'. In the maturation there is the development of mellowness and a decrease of harshness. Flavours associated with mature whisky are vanilla, floral, woody, spicy and smooth. The undesirable flavours that dissipate are sour, oily, sulphury and grassy. Various components are extracted from the wood, including those developed by wood charring. The major flavour components of whisky are listed in Table 6.1.

Usually the lighter bodied spirits generated on a continuous still are blended with a range of heavier bodied spirits coming either from batch stills or by distillation to lower ethanol concentrations in column stills. In the decantation process, the various whiskies are decanted into troughs by which they flow

**Table 6.1** Flavour constituents of whisky.

| | |
|---|---|
| *Main congeners* | Ethyl octadecanoate |
| Acetaldehyde | Ethyl octanoate |
| Ethyl acetate | 4-Ethyl guaiacol |
| Isobutanol | 2-Ethylphenol |
| Methanol | 4-Ethylphenol |
| 2-Methyl butanol | Ethyl undecanoate |
| 3-Methyl butanol | Eugenol |
| *n*-Propanol | Furfural |
| | Furfuryl formate |
| *Other congeners* | Gallic acid |
| Acetyl furan | Guaiacol |
| Benzaldehyde | Hexadecanol |
| Butanol | Hexanol |
| Coniferaldehyde | 5-Hydroxymethyl furfural |
| *m-, o-, p*-Cresol | Isoamyl acetate |
| Decanoic acid | Isoamyl alcohol |
| Decanol | Isoamyl decanoate |
| Diethoxypropane | Isoamyl octanoate |
| Diethyl succinate | *Cis*-Oak lactone |
| Dimethyl disulphide | *Trans*-Oak lactone |
| Dimethyl sulphide | Octanol |
| Dimethyl trisulphide | Phenol |
| Dodecanoic acid | Phenylethanol |
| Dodecanol | Phenylethyl acetate |
| Ellagic acid | Phenylethyl butanoate |
| 3-Ethoxypropanal | Scopoletin |
| Ethyl butanoate | Synapaldehyde |
| Ethyl decanoate | Syringealdehyde |
| Ethyl dodecanoate | Syringic acid |
| Ethyl hexadecanoate | Tetradecanoic acid |
| Ethyl hexadecenoate | Triethoxypropane |
| Ethyl hexanoate | Vanillic acid |
| Ethyl lactate | Vanillin |
| Ethyl nonanoate | 3,5-Xylenol |

to a blending vat wherein they are mixed by mechanical agitator and compressed air. Then 'de-proofing water' is added to take the product to its final strength.

In Scotland, the final products may be a blend of whiskies from more than ten grain distilleries and up to a hundred malt distilleries. There is an astonishing interaction and cooperation between separate companies to enable this. The blending is deliberately complex so that the unavailability of one or two whiskies in any single blending will not be noticeable. In other countries where there are far fewer distilleries, batch-to-batch variation must be achieved by varying conditions within the distilleries themselves – for example, the grist or the fermentation and distillation conditions.

Most whisky is filtered. Insoluble fractions, notably lignins and long chain esters of fatty acids, are removed by cooling to as low as −10°C and filtration, typically in plate and frame devices with diatomaceous earth as filter aid.

### Whiskey variants

*Bourbon* (United States) is made principally from corn (maize) plus added rye and barley and is aged in charred barrels. A close relative is *Tennessee whiskey* (United States), which is produced using a sour mash process. *Canadian whisky* (Canada) is a light product from rye and malted rye, with some corn and malted barley. *Corn whiskey* (United States) is from maize and is aged in barrels that have not been charred. *Rye whiskey* (United States) is from rye mixed with corn and barley and is aged in newly charred oak barrels.

## Cognac

The grape vines employed for the base wine for cognac production are nearly all from Charente and the adjacent regions of Deux-Sèvres and Dordogne. Furthermore, the grape varieties must be either Ugni blanc, Colmbard or Folle Blanc, with the exception that up to 10% can be wines from Jurançon blanc, Semillon, Montils, Blanc ramé or Select.

Ugni blanc is by far the major variety, affording wine high in acidity and relatively low in alcohol which renders it most suitable for distillation. The reader is referred to Cantagrel and Galy (2003) for details of the wine making intricacies. But suffice to say here that the microflora employed for fermentation is endogenous, with one report suggesting that more than 650 yeasts are involved. The belief is that active dry yeast (ADY) leads to the production of inferior products. Sulphur dioxide is not employed. Fermentation is relatively fast, the wines being maintained on the lees and subjected to malo-lactic fermentation. The sooner the distillation after fermentation, the better the quality of the product as there is less development of ethyl butyrate and acrolein (from the decomposition of glycerol, see Chapter 5).

The distillation employed in the production of cognac is known as the Charente process. The still must have a capacity of less than 30 hL, which means that the maximum practical working volume is no more than 25 hL. The vessel must be heated by an open fire.

Two successive distillations yield a spirit of <72% ABV. In the first stage, 27–30% ABV is achieved. In the further distillation of this, three major fractions are generated: the heads, the heart (spirit cut) and the seconds. The heads, comprising 1–2% of the total, contain the most volatile components and are considered detrimental. The most 'noble' components are in the heart and herein is the cognac spirit to be matured. The seconds are recycled.

The nature of the wood employed for ageing of cognac has great significance. The fineness of the grain impacts the extent to which phenolics and other tannins are extracted, as does the shape and size of the barrel made from that wood and the extent to which the wood is charred in the shaping process. The wood is generally dried in the open air for over 3 years. New spirit is introduced to this new wood for a period of 8–12 months before transferring

**Table 6.2**   Changes in volatiles in cognac during different periods of ageing in wood.

| Component | Concentration (mgL$^{-1}$) | | |
|---|---|---|---|
| | 0.7 years | 5 years | 13 years |
| Coniferaldehyde | 3.7 | 5.9 | 6.7 |
| Gallic acid | 4.6 | 9.0 | 15.3 |
| Sinapaldehyde | 9.5 | 17.8 | 17.0 |
| Syringaldehyde | 2.3 | 8.9 | 17.6 |
| Syringic acid | 0.6 | 2.6 | 7.0 |
| Vanillic acid | 0.3 | 1.4 | 2.8 |
| Vanillin | 0.9 | 4.4 | 8.8 |

Derived from Cantagrel (2003).

to older barrels, thereby avoiding the pick up and development of excessive astringent and bitter characteristics. Oxygen enters through the stave and is used by enzymes contributed by moulds in reactions that have a role in the ageing process. There is also volatile loss through the stave. The changes in key wood-derived volatiles that result from different periods of ageing are depicted in Table 6.2.

Several batches will be blended during ageing. New distillates at 70% ABV are lowered in successive stages to the 40% ABV level at which the product is bottled.

## Armagnac and wine spirits

Armagnac is in South West France. The three main vinestocks used for armagnac are as for cognac, with Ugni blanc again being preferred on account of a reduced risk of rot as it comes to maturity rapidly. The wines must be distilled in the Appellation area, with the maximum content of distilled alcohol allowed being 72% ABV. Again, the use of sulphur dioxide is forbidden.

Two types of still are used: the continuous Armagnac still and two-stage pot stills. Continuous armagnac stills are fabricated from copper and are operated as described by Bertrand (2003a). Operational variables are the rate of wine flow and the heating regimen. Heating is always by open fire, although nowadays it will probably be fuelled by propane gas rather than by wood. Just as for whisky, the three components emerging from a still are heads, body and tailings.

The two-stage pot still is comparable with that used for cognac.

Wine spirits are usually aged in oak casks. Coarse-grained wood is preferred because more oxygen can then enter to polymerise tannins. Oxygen ingress is also important for the oxidation of some of the alcohol to acetic acid, which in turn reacts with alcohol to generate flavoursome esters during ageing. A comparison of the key analytical parameters for armagnac, cognac and brandy is given in Table 6.3.

**Table 6.3**   An analytical comparison of wine spirits.

| Parameter | Cognac | Armagnac | Brandy |
|---|---|---|---|
| Alcohol (%ABV) | 40.04 | 41.4 | 45.46 |
| Total acidity (as acetic acid) | 103.6 | 153.9 | 31.46 |
| Volatile acidity (as acetic acid) | 59.3 | 106.5 | 19.06 |
| Aldehydes | 19.3 | 23.3 | 25.33 |
| Esters | 72.9 | 109.6 | 54.8 |
| Higher alcohols | 444.4 | 441.4 | 258.4 |
| Total volatile substances | 632 | 682.1 | 357.5 |

Based on Bertrand (2003b). Apart from alcohol, units are $g\,h\,L^{-1}$

Expert blending is performed and the alcohol concentration lowered to a minimum of 40% by the addition of distilled water. Caramel may be added to enhance colour. The product is held at $-5°C$ for 1 week prior to filtration through cellulose.

*Brandy* is obtained from wine spirits blended or not with wine distillates distilled to less than 94.8% ABV, such distillates not exceeding 50 proof maximum in the final product. The product is aged in oak for more than 1 year, unless the casks hold less than 1000 L in which case ageing must be for a minimum of 6 months. According to Bertrand (2003), the making of brandy is an opportunity to salvage defective wines or deal with production surpluses, although top quality brandies may be made from wine specifically produced for the purpose. Brandies must be >37.5% ABV.

# Rum

Rum primarily originated in the Caribbean, although the first references to liqueurs obtained from sugar cane are from India. Sugar cane was introduced to the Caribbean by Christopher Columbus in 1493.

The chief producing countries are Barbados and Santo Domingo. Nowadays the coastal planes of Guyana (Demerara) are rich in estates producing sugar cane (*Saccharum officinarum*).

At harvest time the fields of sugar cane are set alight in order to sanitise the soil, the stems are scorched in this process and the canes subsequently wither and are harvested by machete, a strategy thought to yield a superior product when compared with rum made from cane harvested by machine.

The canes are topped to remove the leafy parts and the cane then ferried to mills. There is considerable contamination with *Leuconostoc mesenteroides*, which produces a gum that causes problems during extraction. It is important to avoid delays between cutting and milling, and the maximum time elapse should be less than 24 h.

During processing, the canes are cut and crushed and the juice limed, clarified and evaporated. Various fractions are generated, but the key product for rum is molasses. Four to five tons of molasses are typically obtained per 100 tons cane.

The nature of the molasses depends on cane variety, soil type, climate, cultivation and harvesting conditions. They are delivered hot to the distillery either by pipe or by tanker and are stored at 45°C. The molasses are pumped at 85–88°Brix and are mixed with water in line. Lighter flavour rums may incorporate cane juice (12–16% w/v sucrose).

Formerly adventitious yeasts were used to effect fermentation, but nowadays pure cultures of *S. cerevisiae*, *S. bayanus* and *Schizosaccharomyces pombe* are used. They are propagated from slopes by successively scaled up incubations using sucrose as the carbon source.

Prior to fermentation, the molasses are diluted to 45°Brix and their temperature elevated to 70°C in order to destroy contaminating organisms. The pH is lowered by the addition of sulphuric acid and the whole clarified by putting into a conical-bottomed settling tank, from which the sludge can be decanted from the cone. Ammonium sulphate is added as a source of nitrogen.

Fermentation is conducted at 30–33°C in cylindroconical vessels that may be closed or open. The final sugar content will be 16–20°Brix and this is reached in 24 h with an alcohol yield of 5–7% ABV. Some high-gravity fermentations nowadays furnish 10–13% ABV.

Distillation is conducted in pot stills that were traditionally of copper or wood but now more likely to be fabricated from stainless steel. As for whisky, there are also column stills of stainless steel or copper (Coffey stills).

Pot stills afford heavier rums that need prolonged maturation, whereas the column stills are employed for lighter rums, or to generate the neutral spirits that can be used for the production of gin and vodka. Distillates are collected at 80–94% ABV for rums and >96% for neutral spirits.

Pot distillation of rum is exactly analogous to the techniques used in the production of whisky. The pot is charged with wash at approximately 5.5% ABV and the retort charged with low wines at 51–52% ABV from the previous distillation. The fractions obtained are heads, spirits, and feints. The heads are rich in esters and are collected for the initial 5 min in the low wines receiver. The ensuing spirits are collected for 1.5–2 h at 85% ABV. When the emerging strength drops to 43% ABV, the flow is again diverted to the low wines receiver in order to collect the feints. Distillation is completed when the distillate approaches some 1% ABV.

Column distillation allows ten times more output than does pot distillation and is performed exactly analogously to the whisky process.

Rum is aged in Bourbon oak casks. It is racked at 83–85% ABV. As the main production locale is tropical, ageing is quite rapid and may be complete within 6 months. There may first have been a blending of light rums produced in column stills with heavier rums out of pots. Furthermore, there may be transfers between casks for successive maturation periods. Finally rum is chilled to −10°C and filtered to remove fatty acid esters prior to dilution to final strength and packaging.

## Bibliography

Bertrand, A. (2003a) Armagnac and wine-spirits. In *Fermented Beverage Production*, 2nd edn. (eds A.G.H. Lea & J.R. Piggott), pp. 213–238. New York: Kluwer/Plenum.

Bertrand, A. (2003b) Armagnac, brandy and cognac and their manufacture. In *Encyclopedia of Food Sciences and Nutrition* (eds B. Caballero, L.C. Trugo & P.M. Finglas), pp. 584–601. Oxford: Academic Press.

Cantagrel, R. (2003) Chemical composition and analysis of cognac. In *Encyclopedia of Food Sciences and Nutrition* (eds B. Caballero, L.C. Trugo & P.M. Finglas), pp. 601–606. Oxford: Academic Press.

Cantagrel, R. & Galy, B. (2003) From vine to cognac. In *Fermented Beverage Production*, 2nd edn. (eds A.G.H. Lea & J.R. Piggott), pp. 195–212. New York: Kluwer/Plenum.

Huetz de Lemps (1997) *Histoire du Rhum*. Paris: Èditions Desjonqueres.

Lafon, J., Couillaud, P. & Gay-Bellile, F. (1973) *Le Cognac, sa Distillation*. Paris: Editions JB Ballière.

Lyons, T.P., Kelsall, D.R. & Murtagh, J.E., eds (1995) *The Alcohol Textbook*. Nottingham University Press.

Moss, M.S. & Hume, J.R. (1981) *The Making of Scotch Whisky*. Ashnurton: James and James.

Nicol, D.A. (2003) *Rum*. In *Fermented Beverage Production*, 2nd edn. (eds A.G.H. Lea, & J.R. Piggott), pp. 263–287. New York: Kluwer/Plenum.

Piggott, J.R., ed. (1983) *Flavour of Distilled Beverages: Origin and Development*. Chichester: Ellis Horwood.

Piggott, J.R., Sharp, R. & Duncan, R.E.B. (1989) *The Science and Technology of Whiskies*. Harlow: Longman.

Russell, I., ed. (2003) Whisky: technology, production and marketing. In *Handbook of Alcoholic Beverages*, Vol. 1 (eds I. Russell, C.W. Bamforth & G.G. Stewart). London: Academic.

# Chapter 7

# Flavoured Spirits

These products have a base of high purity alcohol, neutral alcohol that has been distilled to a strength in excess of 96% ABV. They are for the most part marketed at 35–40% ABV and do not rely on any maturation period in their production. Many of them are colourless.

Vodka ('little water') is essentially pure alcohol in water, though flavoured variants are available. Gin comprises distilled alcohol flavoured with a range of botanicals. In the same stable come Genever (like gin, flavoured with juniper), Aquavit (caraway and/or dill), Anis (aniseed, star anise, fennel) and Ouzo (aniseed, mastic).

## Vodka

Vodka comprises pure unaged spirit distilled from alcoholic matrices of various origins and usually filtered through charcoal. It is defined in the EU as a:

> spirit drink produced by either rectifying ethyl alcohol of agricultural origin or filtering it through activated charcoal . . .

The EU defined the characteristics of neutral alcohol ('Ethyl alcohol of agricultural origin for use in blending alcoholic beverages') according to Council Regulation No. 1576/89 (Table 7.1).

Materials added in the production of vodka include sugar at up to $2 \, \text{g} \, \text{L}^{-1}$ and citric acid at up to $150 \, \text{mg} \, \text{L}^{-1}$. Some vodkas have glycerol or propylene glycol added to enhance the mouthfeel. Amongst the flavoured vodkas are ones infused with pepper, a Polish product in which buffalo grass is steeped in the spirit and a Russian variant in which the vodka is treated with apple and pear tree leaves, brandy and port.

The neutral alcohol base is frequently produced quite separately from the vodka *per se*, perhaps by a different company. It is chiefly produced from cereals (e.g. corn, wheat) but other sources of fermentable carbohydrate include beet and molasses in Western countries, cane sugar in South America and Africa, and potatoes in Poland and Russia.

The fermentation is, of course, effected by *Saccharomyces cerevisiae*, notably distillers' strains.

The alcohol is purified and concentrated by continuous stills with 2–5 columns. The first of these is a 'wash column' that separates alcohol from

**Table 7.1**  Characteristics of neutral alcohol according to Council Regulation No. 1576/89.

| | |
|---|---|
| Organoleptic characteristics | No detectable taste other than that of the raw material |
| Minimum alcoholic strength by volume | 96% vol. |
| Maximum values of residue elements | |
| *Total acidity*: Expressed in g of acetic acid per hl of alcohol at 100% vol. | 1.5 (15 ppm) |
| *Esters*: Expressed in g of ethyl acetate per hl of alcohol at 100% vol. | 1.3 (13 ppm) |
| *Aldehydes*: Expressed in g of acetaldehyde per hl of alcohol at 100% vol. | 0.5 (5 ppm) |
| *Higher alcohols*: Expressed in g of 2-methyl, 1-propanol (iso-butanol) per hl of alcohol at 100% vol. | 0.5 (5 ppm) |
| *Methanol*: Expressed in g per hl of alcohol at 100% vol. | 50 (500 ppm) |
| *Dry extract*: Expressed in g per hl of alcohol at 100% vol. | 1.5 (15 ppm) |
| *Volatile bases containing nitrogen*: Expressed in g of nitrogen per hl of alcohol at 100% vol. | 0.1 (1 ppm) |
| *Furfural* | Not detectable |

Data from http://www.distill.com/specs/EU.html

the wash. The second major column is the 'rectifier' in which alcohol is concentrated. There may be a 'purifier' between the wash column and the final rectifier.

The wash column distillate is introduced halfway up the extractive distillation column and water (approximately 20 times more than wash) is fed in at the top. This procedure impacts the volatilisation of components of the wash and encourages the removal of volatiles. Ethanol mostly leaves with water at the base of the column, prior to concentration in the final rectification column.

Treatment with activated carbon is either by using a dispersion of purified charcoal in a tank prior to its removal by filtration or by passing the spirit through columns that contain charcoal in granular form.

# Gin

The word gin is a corruption of *genievre*, the French word for juniper. Distilled gin is produced by distilling neutral alcohol and water in the presence of botanicals, of which juniper, coriander and angelica are key. The product is diluted further with alcohol and finally brought to its final strength with water.

In the EU, a drink can be called gin if it is produced by addition to ethanol (of agricultural origin) natural (or nature-identical) flavourants such that the taste is predominantly one of juniper. 'Compounded gin' is made by adding essences to ethanol and this can not be called gin.

The alcohol for gin may come from grain-, molasses-, potato-, grape- or whey-based fermentations.

The prime traditional flavourants are the juniper berry (*Juniperus communis*), coriander seed (*Coriandrum sativum*) and Angelica (*Archangelicum officinalis*), together with the peel of orange and lemon.

Other materials may also be used in the formulation of gins and these include cassia bark, cubeb beris, liquorice, orris, almonds and grains of paradise.

Water quality is critical for the production of gin and, as for beer, this explains the traditional locales where the drink was first made and became popular. These days, as for beer, water purification and salt adjustment protocols mean that the production region is of no significance.

Gin is produced in copper pot stills similar to those used in the production of whisky. Nowadays they tend to be steam-heated rather than direct fired. The still is charged with water prior to adding alcohol to the desired concentration which is typically 60% ABV. The botanics are added either loose or suspended in a bag. The still is closed and heated.

The 'heads' emerge first, followed by the main fraction, of some 80% ABV, which is collected as gin. The 'tails' comprise the later fractions in which alcohol concentration is falling. They are collected with maximum heating and are combined with the heads as 'feints' to be purified in a separate distillation or alternatively sent to the alcohol supplier.

Sloe gin is produced by steeping berries of the sloe (*Prunus spinosa*) in gin. The mix is sweetened with sugar, filtered and bottled. Nowadays flavourants may be employed in place of the berries *per se*.

Pimms is based on a secret recipe and is compounded from gin and liqueurs.

## Liqueurs

These are produced by dissolving or blending several components. For the most part, they are 35–45% ABV, although some are less strong.

The definition of a liqueur (and indeed other alcoholic beverages) is through European Council regulation 1576/89 (Table 7.2).

The alcohol must not be synthetic (i.e. derived from petroleum), but rather must be from a fermentation process. The other key ingredients in these products are sugar (to deliver both sweetness and mouthfeel), flavours (that may be either the plant material *per se* or distilled essential oils or extracts from those botanics) and colour (which again may be of 'natural' origin or via an approved colourant).

Cream liqueurs incorporate milk fat, sodium caseinate and an emulsifier. Through homogenisation procedures, the size of the fat globules is reduced to one that allows a stable emulsion to be obtained.

A representative list of liqueurs is offered in Table 7.3.

**Table 7.2**  EU definitions of categories of alcoholic beverages – Council Regulation 1576/89; Article 1, Section 4.

*A. Rum*

(1) A spirit drink produced exclusively by alcoholic fermentation and distillation, either from molasses or syrup produced in the manufacture of cane sugar or from sugar cane juice itself, and distilled at less than 96% vol., so that the distillate has the discernible specific organoleptic characteristics of rum

(2) The spirit produced exclusively by alcoholic fermentation and distillation of sugar cane juice, which has the aromatic characteristics specific to rum, and a content of volatile substances equal to or exceeding $225\,g\,hl^{-1}$ of alcohol of 100% vol. (2250 ppm). This spirit may be marketed with the word 'agricultural' qualifying the designation 'rum' accompanied by any of the geographical designation of the French Overseas Departments as listed in Annex II

(3) Bottled at a minimum alcoholic strength of 37.5% v/v

*B. Whisky or whiskey*

(1) A spirit drink produced by the distillation of a mash of cereals

- saccharified by the diastase of the malt contained therein, with or without other natural enzymes
- fermented by the action of yeast
- distilled at less than 94.8% vol, so that the distillate has an aroma and taste derived from the raw materials used
- and matured for at least 3 years in wooden casks not exceeding 700 L capacity

(2) Bottled at a minimum alcoholic strength of 40% v/v

*C. Grain spirit*

(1) A spirit drink produced by the distillation of a fermented mash of cereals, and having organoleptic characteristics derived from the raw materials used 'Grain Spirit' may be replaced by 'Korn' or 'Kornbrand', for the drink produced in Germany and in regions of the Community where German is one of the official languages, provided that this drink is traditionally produced in these regions, and if the grain spirit is obtained there without any additive:

- either exclusively by the distillation of a fermented mash of whole grain of wheat, barley, oats, rye or buckwheat with all their component parts
- or by the redistillation of a distillate obtained in accordance with the first subparagraph

(2) For a grain spirit to be designated 'grain brandy', it must have been obtained by distillation at less than 95% vol. from a fermented mash of cereals, presenting organoleptic features deriving from the raw materials used

(3) Bottled at a minimum alcoholic strength of 35% v/v

*D. Wine spirit*

(1) A spirit drink

- produced exclusively by the distillation at less than 86% vol., of wine or wine fortified for distillation, or by the redistillation of a wine distillate at less than 86% vol.
- containing a quantity of volatile substances equal to or exceeding $125\,g\,hl^{-1}$ of 100% vol. alcohol (1250 ppm), and
- having a maximum methyl alcohol content of $200\,g\,hl^{-1}$ of 100% vol. alcohol (2000 ppm)

Where this drink has been matured, it may continue to be marketed as 'wine spirit' if it has matured for as long as, or longer than, the period stipulated for the product referred to in (E)

(2) Bottled at a minimum alcoholic strength of 37.5% v/v

**Table 7.2** *Continued*

*E. Brandy or Weinbrand*

(1) A spirit drink

- produced from wine spirit, whether or not blended with a wine distillate distilled at less than 94.8% vol., provided that the said distillate does not exceed a maximum of 50% by volume of the finished product
- matured for at least 1 year in oak receptables, or for at least 6 months in oak casks with a capacity of less than 1000 L
- containing a quantity of volatile substances equal to or exceeding $125\,\mathrm{g\,hl^{-1}}$ of 100% vol. alcohol (1250 ppm), and derived exclusively from the distillation or redistillation of the raw materials used
- having a maximum methyl alcohol content of $200\,\mathrm{g\,hl^{-1}}$ of 100% vol. alcohol (2000 ppm)

(2) Bottled at a minimum alcoholic strength of 35% v/v

*F. Grape marc spirit or grape marc*

(1) (a) A spirit drink

- produced from grape marc fermented and distilled either directly by water vapour, or after water has been added. A percentage of lees that is to be determined in accordance with the procedure laid down in Article 15 may be added to the marc, the distillation being carried out in the presence of the marc itself at less than 86% vol. Redistillation at the same alcoholic strength is authorised
- containing a quantity of volatile substances equal to or exceeding $140\,\mathrm{g\,hl^{-1}}$ of 100% vol. alcohol (1400 ppm), and having a maximum methyl alcohol content of $1000\,\mathrm{g\,hl^{-1}}$ of 100% vol. alcohol (10 000 ppm)

(b) However, during the transitional period provided for Portugal in the 1985 Act of Accession, subparagraph (a) shall not preclude the marketing in Portugal of grape marc spirit produced in Portugal and having a maximum methyl alcohol content of $1500\,\mathrm{g\,hl^{-1}}$ of 100% vol. (15 000 ppm)

(2) The name 'grape marc' or 'grape marc spirit' may be replaced by the designation 'grappa' solely for the spirit drink produced in Italy

(3) Bottled at a minimum alcoholic strength of 37.5% v/v

*G. Fruit marc spirit*

(1) A spirit drink produced by the fermentation and distillation of fruit marc. The distillation conditions, product characteristics and other provisions shall be established in accordance with the procedure laid down in Article 15

(2) Bottled at a minimum alcoholic strength of 37.5% v/v

*H. Raisin spirit or raisin brandy*

(1) A spirit drink produced by the distillation of the product obtained by the alcoholic fermentation of extract of dried grapes of the 'Corinth Black' or 'Malaga Muscat' varieties, distilled at less than 94.5% vol., so that the distillate has an aroma and taste derived from the raw materials used

(2) Bottled at a minimum alcoholic strength of 37.5% v/v

*I. Fruit spirits*

(1) (a) Spirit drinks

- produced exclusively by the alcoholic fermentation and distillation of fleshy fruit or must of such fruit, with or without stones
- distilled at less than 86% vol., so that the distillate has an aroma and taste derived from the fruits distilled
- having a quantity of volatile substances equal to or exceeding $200\,\mathrm{g\,hl^{-1}}$ of 100% vol. alcohol (2000 ppm)
- having a maximum methyl alcohol content of $1000\,\mathrm{g\,hl^{-1}}$ of 100% vol. alcohol (10 000 ppm), and
- in the case of stone-fruit spirits, having a hydrocyanic acid content not exceeding $10\,\mathrm{g\,hl^{-1}}$ vol. alcohol (100 ppm)

**Table 7.2**    *Continued*

(b) Drinks thus defined shall be called 'spirit' preceded by the name of the fruit, such as cherry spirit or kirsch, plum spirit or slivovitz, mirabelle, peach, apple, pear, apricot, fig, citrus or grape spirit or other fruit spirits. They may also be called 'wasser' with the name of the fruit

The name 'Williams' may be used only to describe pear spirit produced solely from pears of the 'Williams' variety

Whenever two or more fruits are distilled together, the product shall be called 'fruit spirit'. The name may be supplemented by that of each fruit, in decreasing order of quantity used

(c) The cases and conditions in which the name of the fruit may replace the name 'spirit' preceded by the name of the fruit in question shall be determined in accordance with the procedure laid down in Article 15

(2) The name 'spirit' preceded by the name of the fruit may also be used for spirit drinks produced by macerating, within the minimum proportion of 100 kg of fruit per 20 L of 100% vol. alcohol, certain berries and other fruit such as raspberries, blackberries, bilberries and others, whether partially fermented or unfermented, in ethyl alcohol of agricultural origin or in spirit or distillate as defined in this Regulation, followed by distillation

The condition for using the name 'spirit' preceded by the name of the fruit with a view to avoiding confusion with the fruit spirits in point 1 and the fruit in question shall be determined by the procedure laid down in Article 15.

(3) The spirit drinks obtained by macerating unfermented whole fruit such as that referred to in point 2, in ethyl alcohol of agricultural origin, followed by distillation, may be called 'geist', with the name of the fruit

(4) Bottled at a minimum alcoholic strength of 37.5% v/v

*J. Cider spirit, cider brandy or perry spirit*

(1) Spirit drinks

- produced exclusively by the distillation of cider or perry, and
- satisfying the requirements of the second, third and fourth indents of subparagraph (I) (1) (a) relating to fruit spirits

(2) Bottled at a minimum alcoholic strength of 37.5% v/v

*K. Gentian spirit*

(1) A spirit drink produced from a distillate of gentian, itself obtained by the fermentation of gentian roots with or without the addition of ethyl alcohol of agricultural origin

(2) Bottled at a minimum alcoholic strength of 37.5% v/v

*L. Fruit spirit drinks*

(1) Spirit drinks obtained by macerating fruit in ethyl alcohol of agricultural origin and/or in distillate of agricultural origin and/or in spirits as defined in this Regulation and within a minimum proportion to be determined by means of the procedure laid down in Article 15

The flavouring of these spirit drinks may be supplemented by flavouring substances and/or flavouring preparations other than those which come from the fruit used. These flavouring substances and flavouring preparations are defined respectively in Article 1 (2) (b) (i) and (c) of Directive 88/388/EEC. However, the characteristic taste of the drink and its colour must come exclusively from the fruit used

(2) The drinks so defined shall be called 'spirit drinks' or 'spirit' preceded by the name of the fruit. The cases and conditions in which the name of the fruit may replace those names shall be determined by means of the procedure laid down in Article 15. However, the name 'Pacharan' may be used solely for the 'fruit spirit drink' manufactured in Spain, and obtained by macerating sloes (*Prunus esponisa*) within the minimum proportion of 250 g of fruit per L of pure alcohol

**Table 7.2**   *Continued*

    (3) Bottled at a minimum alcoholic strength of 37.5% v/v

*M. Juniper-flavoured spirit drinks*

    (1) (a) Spirit drinks produced by flavouring ethyl alcohol of agricultural origin and/or grain spirit and/or grain distillate with juniper (*Junipers communis*) berries
Other natural and/or nature-identical flavouring substances as defined in Article 1 (2) (b) (i) and (ii) of Directive 88/388/EEC and/or flavouring preparations defined in Article 1 (2) (c) of that Directive, and/or aromatic plants or parts of aromatic plants may be used in addition, but the organoleptic characteristics of juniper must be discernible, even if they are sometimes attenuated

        (b) The drinks may be called 'Wacholder', 'ginebra' or 'genebra'. Use of these names is to be determined in accordance with the procedure laid down in Article 15

        (c) The alcohols used for the spirit drinks called 'genievre', 'jenever', 'genever' and 'peket' must be organoleptically suitable for the manufacture of the aforementioned products, and have a maximum methyl content of $5\,\mathrm{g\,hl^{-1}}$ of 100% vol. alcohol (50 ppm), and a maximum aldehyde content expressed as acetaldehyde of $0.2\,\mathrm{g\,hl^{-1}}$ of 100% vol. alcohol (2 ppm). In the case of such products, the taste of juniper berries need not be discernible

    (2) (a) The drink may be called 'gin' if it is produced by flavouring organoleptically suitable ethyl alcohol of agricultural origin with natural and/or nature-identical flavouring substances as defined in Article 1 (2) (b) (i) and (ii) of Directive 88/388/EEC and/or flavouring preparations as defined in Article 1 (2) (c) of that Directive so that the taste is predominantly that of juniper

        (b) The drink may be called 'distilled gin', if it is produced solely by redistilling organoleptically suitable ethyl alcohol of agricultural origin of an appropriate quality, with an initial alcoholic strength of at least 96% vol., in stills traditionally used for gin, in the presence of juniper berries and of other natural botanicals, provided that the juniper taste is predominant. The term 'distilled gin' may also apply to a mixture of the product of such distillation and ethyl alcohol of agricultural origin with the same composition, purity and alcoholic strength. Natural and/or nature-identical flavouring substances and/or flavouring preparations as specified at (a) may also be used to flavour distilled gin. 'London Gin' is a type of distilled gin
Gin obtained simply by adding essences or flavouring to ethyl alcohol of agricultural origin shall not qualify for the description 'distilled gin'

    (3) Bottled at a minimum alcoholic strength of 37.5% v/v

*N. Caraway-flavoured spirit drinks*

    (1) Spirit drinks produced by flavouring ethyl alcohol of agricultural origin with caraway (*Carum carvi* L.)
Other natural and/or nature-identical flavouring preparations as defined in Article 1 (2) (b) (i) and (ii) of Directive 88/388/EEC, and/or flavouring substances as defined in Article 1 (2) (c) of that Directive, may additionally be used but there must be a predominant taste of caraway

    (2) (a) The spirit drinks defined in point 1 may also be called 'akvavit' or 'aquavit', if they are flavoured with a distillate of plants or spices
Other flavouring substances specified in the second subparagraph of point 1 may be used in addition, but the flavour of these drinks is largely attributable to distillates of caraway and/or dill (*Anethum graveolens L.*) seeds, the use of essential oils being prohibited

        (b) The bitter substances must not obviously dominate the taste; the dry extract content may not exceed 1.5 g per 100 ml

    (3) Bottled at a minimum alcoholic strength of 30% v/v, except akvavit which is bottled at a minimum alcoholic strength of 37.5% v/v

**Table 7.2**    *Continued*

*O. Aniseed-flavoured spirit drinks*

(1) Spirit drinks produced by flavouring ethyl alcohol of agricultural origin with natural extracts of star anise (*Illicium verum*), anise (*Pimpinella anisum*), fennel (*Foeniculum vulgare*), or any other plant, which contains the same principal aromatic constituent, using one of the following processes:

- maceration and/or distillation;
- redistillation of the alcohol in the presence of the seeds or other parts of the plants specified above
- addition of natural distilled extracts of aniseed-flavoured plants
- a combination of these three methods

Other natural plant extracts or aromatic seeds may also be used, but the aniseed taste must remain predominant

(2) For an aniseed-flavoured spirit drink to be called 'pastis', it must also contain natural extracts of liquorice root (*Glycyrrhiza glabra*), which implies the presence of the colourants known as 'chalcones' as well as glycyrrhizic acid, the minimum and maximum levels of which must be 0.05 and $0.5 \, g \, L^{-1}$ grams per litre respectively

Pastis contains less than 100 g of sugar per L and has a minimum and maximum anethole level of 1.5 and $2 \, g \, L^{-1}$ respectively

(3) For an aniseed-flavoured spirit drink to be called 'ouzo', it must:

- have been produced exclusively in Greece
- have been produced by blending alcohols flavoured by means of distillation or maceration, using aniseed and possibly fennel seed, mastic from a lentiseus indigenous to the island of Chios (*Pistacia lentiscus* Chia or latifolia) and other aromatic seeds, plants and fruits. The alcohol flavoured by distillation must represent at least 20% of the alcoholic strength of the ouzo

That distillate must:

- have been produced by distillation in traditional discontinuous copper stills with a capacity of 1000 L or less
- have an alcoholic strength of not less than 55% vol. and not more than 80% vol. Ouzo must be colourless and have a sugar content of 50 g or less per litre

(4) For an aniseed-flavoured spirit drink to be called 'anis', its characteristic flavour must be derived exclusively from anise (*Pimpinella anisum*) and/or star anise (*Illicium verum*) and/or fennel (*Foeniculum vulgare*). The name 'distilled anis' may be used if the drink contains alcohol distilled in the presence of such seeds, provided such alcohol constitutes at least 20% of the drink's alcoholic strength

(5) Bottled at a minimum alcoholic strength of 15% v/v, except pastis (40% v/v), ouzo (37.5% v/v) and anis (35% v/v)

*P. Bitter-tasting spirit drinks or bitter*

(1) Spirit drinks with a predominantly bitter taste produced by flavouring ethyl alcohol of agricultural origin with natural and/or nature-identical flavouring substances, is defined in Article 1 (2) (b) (i) and (ii) of Directive 88/388/EEC and/or flavouring preparations as defined in Article 1 (2) (c) of that Directive

The drink may also be marketed as 'amer' or 'bitter' with or without another term. This provision shall not affect the possible use of the terms 'amer' or 'bitter' for products not covered by this Article

(2) Bottled at a minimum alcoholic strength of 15% v/v

*Q. Vodka*

(1) A spirit drink produced by either rectifying ethyl alcohol of agricultural origin, or filtering it through activated charcoal, possibly followed by straightforward distillation or an equivalent treatment, so that the organoleptic characteristics of the raw materials used are selectively reduced. The product may be given special organoleptic characteristics, such as a mellow taste, by the addition of flavouring

(2) Bottled at a minimum alcoholic strength of 37.5% v/v

**Table 7.2**   *Continued*

*R. Liqueur*
(1) A spirit drink:
- having a minimum sugar content of $100 \, \text{g L}^{-1}$ expressed as invert sugar, without the prejudice to a different decision taken in accordance with the procedure laid down in Article 15
- produced by flavouring ethyl alcohol of agricultural origin, or a distillate of agricultural origin, or one or more spirit drinks as defined in this Regulation, or a mixture of the above, sweetened and possibly with the addition of products of agricultural origin such as cream, milk or other milk products, fruit, wine, or flavoured wine
(2) The name 'crème de' followed by the name of a fruit or the raw material used, excluding milk products, shall be reserved for liqueurs with a minimum sugar content of $250 \, \text{g L}^{-1}$ expressed as invert sugar
The name 'crème de cassis' shall, however, be reserved for blackcurrant liqueurs containing at least 400 g of sugar, expressed as invert sugar, per L
(3) Bottled at a minimum alcoholic strength of 15% v/v

*S. Egg liqueur/advocaat/avocat/advokat*
(1) A spirit drink whether or not flavoured, obtained from ethyl alcohol of agricultural origin, the ingredients of which are quality egg yolk, egg white and sugar or honey. The minimum sugar or honey content must be $150 \, \text{g L}^{-1}$. The minimum egg yolk content must be $140 \, \text{g L}^{-1}$ of the final product
(2) Bottled at a minimum alcoholic strength of 15% v/v

*T. Liqueur with egg*
(1) A spirit drink whether or not flavoured, obtained from ethyl alcohol of agricultural origin, the ingredients of which are quality egg yolk, egg white and sugar or honey. The minimum sugar or honey content must be $150 \, \text{g L}^{-1}$. The minimum egg yolk content must be $70 \, \text{g L}^{-1}$ of the final product
(2) Bottled at a minimum alcoholic strength of 15% v/v

National provisions may set a minimum alcoholic strength by volume which is higher than the values indicated above. The minimum bottling strengths are taken from Article 3 of this Regulation. Date from http://www.distill.com/specs/EU3.html

**Table 7.3**   Some liqueurs and speciality alcoholic products.

| Product | Notes | Country of origin |
|---------|-------|-------------------|
| Absinthe | Brandy flavoured with sweet almonds and apricots | France |
| Advocaat | Brandy-base. Egg yolks, sugar and vanilla | Holland |
| Amaretto | Apricot kernel and bitter almond flavour | Italy |
| Anis | Anise/star anise/fennel flavour | Diverse |
| Arrack | Distillation of alcohol from grapes, sugar cane, rice or dates. Word means 'sweat' | Arabic |
| Bailey's | Irish Whiskey and chocolate | Ireland |
| Benedictine | Brandy flavoured with 27 plants (including cardamom, cinnamon, cloves, juniper, nutmeg, tea, myrrh) and sugar. Coloured using saffron and caramel | France |
| Campari | Red product made by blending 68 herbs with quinine, Chinese rhubarb, cinchona bark and orange peels | Italy |

**Table 7.3**  *Continued*

| | | |
|---|---|---|
| Cassis | Macerated blackcurrants in neutral spirits and brandy | France |
| Chartreuse | Blend of 130 herbs and honey in brandy | |
| Cherry Brandy | Distilled juice of cherries, fermented in presence of crushed cherry stones, perhaps blended with Armagnac | Mainland Europe |
| Cointreau | Blend of distillates from bitter and sweet orange peel, plus sugar | France |
| Drambuie | Scotch whisky suffused with herbs, spices and heather honey | Scotland |
| Grande Marnier | Cognac blended with distillates of bitter orange and sugar | France |
| Malibu | Light rum/coconut | Barbados |
| Ouzo | Aniseed and fennel and mastic distilled in copper stills <1000 L | Greece |
| Pernod | Spirit base suffused with star anise, fennel, camomile, coriander, veronica and other herbs | France |
| Sambuca | Anis, star anise, elderflower, invert sugar | Italy |
| Southern Comfort | Grain-based spirit containing peach and orange and sugar | United States |
| Tia Maria | Cane spirit/rum base with coffee and spices and sugar | Jamaica |

# Bibliography

Aylott, R.I. (2003) Vodka, gin and other flavored spirits. In *Fermented Beverage Production*, 2nd edn. (eds A.G.H. Lea & J.R. Piggott), pp. 289–308. New York: Kluwer/Plenum.

Begg, O. (1998) *The Vodka Companion*. London: Quinted.

Clutton, D.W. (2003) Liqueurs and speciality products. In *Fermented Beverage Production*, 2nd edn. (eds A.G.H. Lea & J.R. Piggott), pp. 309–334. New York: Kluwer/Plenum.

Coates, G. (2000) *Classic Gin*. London: Prion.

Durkan, A. (1998) *Spirits and Liqueurs*. Lincolnwood: NTC Contemporary.

Hallgarten, P. (1983) *Spirits and Liqueurs*. London: Faber.

Walton, S. (1999) *Complete Guide to Spirits and Liqueurs*. New York: Anness.

# Chapter 8
# Sake

Sake probably emerged from China in the seventh century, although it is claimed that the first rice wine may have been brewed for the emperor in the third century. The first sake was called 'chewing in the mouth sake' on account of its mode of production. Rice was chewed alongside chestnuts or millet and the wad spit into water in a wooden tub where it was allowed to brew for several days. We now know, of course, that the salivary amylase was degrading starch to fermentable sugars that were converted by adventitious yeasts into alcohol. It was a ritualistic process in Shinto festivals.

The advent of sake proper in the Nara period of 710–794 has an origin comparable with that of beer, insofar as rice went mouldy with the consequence of degradation and spontaneous alcoholic fermentation. Part of the rice that had become infected by mould could be saved and used to start a new batch. We now call this *koji*, with the principal micro-organism being *Aspergillus oryzae*.

Through the ages, sake has had profound social and religious significance. Just as for beer or wine, it has served a strong catalytic, functional and social role in the cementing of society.

The Westernisation of Japanese culture, including the fermentation of sake, can be trace to 1853 when Commodore Matthew Perry of the United States Navy arrived in the harbour south of Tokyo. These days there is a fascinating meeting of Western and ancient Eastern cultures in the production of sake.

In 1872, there were more than 30 000 sake breweries in Japan. The Meiji government recognised (as have so many other governments throughout history) that taxation of alcohol production was a useful source of revenue, and the fiscal burden on sakemakers increased annually. By the start of the twentieth century, only 8000 sake brewers survived and the present shape of the industry was established.

The traditional centres for sake production are Nada and Fushimi and great national brands emerge from here (Figs 8.1–8.6). Local brewers produce Jizaki sakes. There is an increasing use of the latest technology, especially by the largest producers, who apply much automation. Modernisation of the industry was greatly aided by the founding in 1904 of the National Research Institute of Brewing, which was started by the Treasury to test sakes.

A shortage of rice during the last Great War obliged sakemakers to supplement the traditional process stream by the addition of pure alcohol, or glucose or glutinous rice as adjuncts. Such approaches remain as standard procedures in the manufacture of many sakes.

**Fig. 8.1**    Washing and steeping of polished rice.

**Fig. 8.2**    Steaming of rice.

**Fig. 8.3**  Making koji.

**Fig. 8.4**  Making sake seed.

**Fig. 8.5**   Feeding steamed rice to the fermenting mash.

**Fig. 8.6**   Filtration of new sake mash.

It was not until 1983 that sake consumption fell to less than 30% of total alcohol consumed. Presently it amounts to about 15% of the total alcohol market. Beer is much more important nowadays, but other competitors include spirits and schochu, which resembles vodka.

There is renewed interest in sake, however, with its perception as a 'natural food'. In 1975, the Japan Sake Brewers Association established labelling practices that led directly to a reduced use of non-traditional ingredients.

## Sake brewing

The brewing of sake retains much ritual and tradition. The Master Brewers (the toji) go about their tasks in the kura, brewing in the coldest months of the year. The toji are an elite breed of artisan that can trace their origins back to the Edo period. They develop their knowledge and stature over many years of practical experience, starting with the most menial of tasks and enduring long hours of heavy manual labour.

The brewers lived in kura for all the 100 days of the brewing season in past times and were forbidden to leave the establishment until after the final mash had begun. Nowadays machines are employed for the heaviest work and the toji work alongside university-trained technicians.

The key ingredients of sake are water and rice. Some 25 kL of water are used for each ton of rice.

The water should be colourless, tasteless and odourless and should contain only traces of minerals and organic components. Again as for beer and gin, sake making sprang from locations where the water was highly prized. For sake this was the Miyamizu water from Nishinomiya, a port in Nada. The water here actually emerges from three sources: subterranean water from the local river, an adjacent mountain, and seawater. The waters mix below a thick layer of fossilised shells and are filtered through it as the stream rises to the surface. The mountain water is rich in carbonates, phosphate and potassium. It also contains much iron, but this is not a problem because it is oxidised by acids in the river water. These days Miyamizu tends to be produced synthetically and is further refined by filtration and aeration.

The rice employed for sake production is the short grain japonica variety that becomes sticky when cooked (Fig. 8.7). It is polished more than is customarily the case for food use. Fifteen per cent of the material (the outer layers) is removed to take down the levels of protein, lipid and minerals that would jeopardise clarity.

Rice is either grown by the brewer or is purchased under subcontract. Two-thirds of the rice is yamadanishika, which originated in the Hyogo prefecture. Breeding has led to greatly improved yields and agronomic characteristics in the rice varieties that are available.

The basic techniques employed in sake brewing have not changed since the late sixteenth century. The process comprises multiple parallel fermentations.

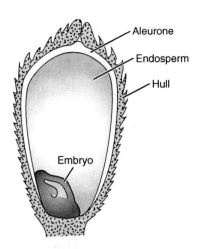

**Fig. 8.7**   A grain of rice.

Saccharification of starch and fermentation of sugar to alcohol occur simultaneously. For the former, koji mould (*A. oryzae*) produces starch-degrading enzymes that generate the fermentable sugar. This is converted by the sake yeast (*Saccharomyces cerevisiae* var. *sake*) to alcohol. The fact that both processes are occurring side by side rather than sequentially means that the yeast does not encounter such a high initial sugar concentration so as to be inhibited. Accordingly the alcohol content achieved can be very high – perhaps 20% ABV – which is higher than for any other directly fermented beverage. The sake yeast actually tolerates up to 30% ethanol.

In overview, steamed rice and water treated with koji mould are added in three separate stages to a highly concentrated yeast mash (moto). The temperature of the final mash (moromi) is maintained at around 15°C and fermentation is allowed to proceed up to 18 days. Accordingly, the basic sequence is making sake is (1) making koji rice; (2) preparing moto and (3) brewing (tsukuri).

### Polishing, steeping and steaming

White rice with a slightly larger grain size than that generally used for food is reduced in weight by 25–30% (or more than 50% for some premium sakes) by the removal of outer layers. The latter jeopardise clarity and flavour and also impact the manner by which the mould grows. The more the polishing undertaken, the cleaner the sake.

The grain is then steeped in water until it reaches around 30% moisture and is then transferred to a large wooden tub (koshiki) with holes in the bottom that admit steam. The mix is placed over a metal tub containing boiling water. This sterilises and gelatinises the rice, rendering it susceptible to the action of koji.

After 50–60 min the rice is removed, divided and cooled depending on which stage in the brewing process it is going to be used in.

## Making koji

Koji comprises *A. oryzae*, which furnishes the necessary hydrolytic enzymes ($\alpha$-glucosidase, glucoamylase, transglucosidase, acid protease, carboxypeptidase) for digesting the starch and the protein. The nature of the process is such that organisms other than the sake yeast will also develop. These include film-forming yeast, micrococci, bacilli and lactic acid bacteria. The rice employed for koji is more refined than the bulk of steamed rice. After steaming, one-fifth of the rice is removed from the koshiki and cooled to about 30°C. It is transferred to a double-walled solar-like room that retains heat. Dried spores of *A. oryzae* are scattered over the surface and kneaded in. Several hours later, the mix is transferred to shallow Japanese cedar wood trays (45 cm × 30 cm × 5.1 cm) that are put on shelves and covered with a cloth. As the koji mould grows, the temperature rises, so the mix is stirred twice every 4 h. After 40–45 h, the boxes are removed and advantage is taken of the low temperatures outside to stop the growth of koji. After cooling, the koji mix is light, dry and flaky and has a distinct aroma of horse chestnuts.

## Making moto

The koji rice for making moto starter is basically treated in the same manner; however, the process is prolonged in order that even higher levels of enzymes are produced. Moto is the seed mash and represents less than 10% of the total rice.

The longest standing method of moto production is mizu moto (bodai moto). Three kilograms of steamed rice already adventitiously infected with yeast from the air is sealed in a cloth bag and buried within uncooked polished rice (87 kg) to which is added 130 L of water. After 4–5 days, the water becomes distinctly cloudy and bubbly and is sour. It is removed by filtration and the polished rice is steamed. A second mash is then produced with this yeasty water, all of the steamed rice and a further 40 kg of koji rice. The moto is ready for use after 5 days.

The disadvantage of this procedure is the emergence of high levels of lactic acid bacteria, causing the ensuing sake to be sour.

Since the 1920s the kimoto method has become the main approach to making moto. The mix comprises 75 kg steamed rice, 30 kg koji rice and 108 L of water. This is divided in the early evening into 16 shallow wooden tubs, each of 70 cm diameter. Toji stir the mixture every 3–4 h through the night (cooling by ambient chill air) and grind the moto the next day using long bamboo poles to which wooden panels are attached. The rice is rubbed against the bottom of the wooden tubs until the grains are reduced to approximately a third of their size and the mash comprises a thick paste. This procedure accelerates the activity of the koji.

The paste is transferred to a single large wooden vat and left for 2–3 days at 8°C. Then buckets of hot water are dropped into the mash, thereby raising the temperature and stimulating airborne yeasts into fermentation. The mix is maintained at 25°C and 20–25 days later, it is used as a starter for the main mash.

It is understood that in the early stages of the process, lactic acid bacteria prevent the growth of other, less desirable organisms. Later on the alcohol developed by yeast kills the lactic acid bacteria and any unwanted wild yeast.

Two other methods have evolved for making moto. The Yamahi process has the same principles as above, but there is an initial mixing of pure koji rice with water so as to accelerate saccharification before the addition of steamed rice. This has become the most popular method. The Sokujo process again has the same basic principle as for raw moto, but here the koji rice is mixed with water and lactic acid added to 5%. At the same time, a pure culture of sake yeast is added to seed the fermentation. Steamed rice is mixed in before cooling and leaving for 2–3 days. Dakitaru is used to raise the temperature to 20°C. After 10–15 days, the mash is ready to use as a starter for the main mash.

### Moromi

After the koji and moto are prepared, they are mixed over 4 days. This is traditionally in large wooden vats (7–20 kL). Increasingly large amounts of rice, koji rice and water are added to the moto on the first, third and fourth days. The addition rates (relative to moto) are 1 : 1 on the first day, 2 : 1 on the third day and 4 : 1 on the fourth day. Through the first and second days, the temperature is allowed to rise to 15°C and the whole is left uncovered. The endogenous acidity prevents the growth of spoilage bacteria. On the third day, the temperature is lowered to 9–10°C and this further suppresses infection. After the fourth-day addition, the ensuing 15–18 days represent a challenge for temperature control, unless the facilities are sufficiently modern to incorporate cooling.

Traditional brewers still operate in the winter months, with the use of slatted windows for cooling. In modern facilities, brewing can proceed around the year.

After 15–18 days, the mixture is filtered through weighted long narrow cotton sacks over a wooden 'sake boat' (sakafune). The sake trickles through a spigot at the base of the boat. The residual lees are sold for the pickling of vegetables and for use in cooking.

New sake is held for 10 days at a low temperature, during which time glucose and acid levels are enzymically lowered. Then it is pasteurised at 60°C and transferred to sealed vats, traditionally fabricated from Japanese cedar, where it will be held for 6–12 months. This allows a mellowing of the product which starts as being yellow, harsh and smelling of koji. During ageing, characters are developed in the sake from the wood. After ageing there will be a blending ('marrying') followed by dilution with water to a final strength of 15–17% ABV and bottling.

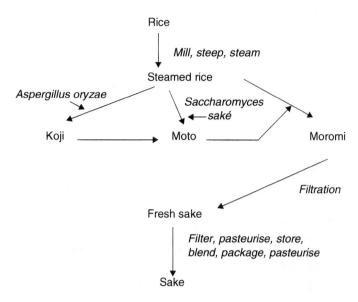

**Fig. 8.8**  Overview of sake production.

## Modern sake making

In modern facilities, the vessels are likely to be fabricated from stainless steel. Rectified alcohol is likely to be employed as a proportion of the sake alcohol, and glucose, lactic acid and monosodium glutamate may also play a role in 'tripling the sake' (cf. earlier). These are added to the final mash as a fourth addition. There is extensive use nowadays of the premier sake yeast strains, with cross-breeding to combine the best properties in a single strain.

In the latest moto processes at high temperature (koon toka mota), the moto mash is raised to 55°C for 5–8 h. Lactic acid is added and the mix is cooled to 20°C prior to the addition of yeast. The entire process takes 5–7 days. It may be computer-controlled. Activated charcoal may be employed in place of sake boats.

A simplified overview of sake production is offered in Fig. 8.8.

## The flavour of sake

Apart from ethanol, significant contributors to the flavour of sake (derived via yeast metabolism) are other alcohols, esters and acids, including lactic acid from the moto stage (Table 8.1).

## Types of sake

Jummai-shu is made from rice alone (reduced to 70% of its original size). Honjozo-shu contains less than 120 L of raw alcohol per ton of white rice

**Table 8.1**  Contributors to the flavour of sake.

| Compound | Typical level (mg L$^{-1}$) |
|---|---|
| Propan-1-ol | 120 |
| Isoamyl alcohol | 70–250 |
| 2-Phenylethanol | 75 |
| Isobutanol | 65 |
| Ethyl acetate | 50–120 |
| Ethyl caproate | 10 |
| Isoamyl acetate | 10 |
| Succinic acid | 500–700 |
| Malic acid | 200–400 |
| Citric acid | 100–500 |
| Acetic acid | 50–200 |
| Lactic acid | 300–500 |

(reduced to 70% of its original size) and the alcohol must be added to the moromi. No glucose is allowed. Ginjo-shu is a special, high-quality variant of Jummai-shu, with the rice reduced to 60% of its original size, no alcohol addition, and very low temperature (10°C) fermentation.

Genshu is undiluted sake (20% ABV) that is served on ice. Taru-zake is cask sake aged in Japanese cypress from the Yoshino region of the Nara prefecture, developing colour and flavour from this wood. Ki-ippon sake is one produced entirely in a single area and not blended with sake originating in other regions. These days it is a name that indicates that the sake is made in a single brewery and that sake must be jummai-shu.

Koshu means old sake, aged for 2–3 years before bottling. As such it contrasts with other sakes that are matured for less than a year and should be drunk young. Nigori-zake has a white and cloudy appearance on account of the use of sacks that do not remove all the particles. Kijo-shu is made by replacing half of the brewing water with sake. Therefore, it is very heavy and sweet (the alcohol suppressing yeast action) and it tends to be used as an aperitif.

Then there are wine-type sakes – rice wine – made with wine yeasts and reaching 13% ABV. Akai-sake is red and made with red koji instead of the customary yellow. As such it is in the realm of gimmick, rather like the inclusion of gold flakes in certain products.

Dry sake is called karakuchi, sweet sake amakuchi.

## Serving temperature

Sakes are customarily served at 20°C when compared for taste. The sake is held in pitchers called tokkuri for pouring into cups known as sakazuki.

The precise manner by which sake is served depends very much on the season, any food that it is accompanying and on the type of sake. Many experts would be of the opinion that warming sake distorts the taste and should be avoided. However, another opinion is that dry sake is better warm (not hot). Nurukan means lukewarm (20–40°C); kan is when sake is 40–45°C (this is standard when sake is asked for warm); atsukan is when the sake is at 55–60°C.

## Bibliography

Inoue, T., Tanaka, J. & Mitsui, S. (1992) Recent advances in Japanese brewing technology. In *Japanese Technology Reviews Section E: Biotechnology*, vol. 2, no. 1. Tokyo: Gordon and Breach.

Kondo, H. (1996) *The Book of Sake*. Tokyo: Kodansha International.

Nunokawa, Y. (1972) Sake. In *Rice Chemistry and Technology* (ed. D.F. Houston), pp. 449–487. St Paul, MN: American Association of Cereal Chemists.

# Chapter 9
# Vinegar

Vinegar is made either by the microbial fermentation of alcohol or by the dilution of acetic acid. It has a pedigree probably spanning more than 10 000 years and, in that time, has been extensively used as food, medicine and for rituals. Wine being the first liquid to have spontaneously soured, we have the derivation of vinegar: *Vin aigre* – in French, sour wine.

Hippocrates understood the medicinal value of vinegar and such uses continued right through the Middle Ages and beyond as an internal and also topical treatment (remember Jack falling down the hill). The acidity represents formidable antimicrobial scope.

Vinegar is nowadays mostly used to afford desired acidic (sour) flavour to foodstuffs and to preserve them. It is still widely produced naturally ('brewed vinegars') by the oxidation of an alcoholic (less than 10–12% ABV) feedstock. The alcohol may be in the form of wine, cider, beer or other alcohol derived from the fermentation of grain, fruit, honey, potatoes, molasses or whey (Table 9.1). In industrial countries, more than 2 L of vinegar are consumed per head each year. Apart from direct use in domestic cooking and in finished foods, it is used extensively *inter alia* for mayonnaises, sauces, ketchups and pickles. For pickling purposes, the acetic acid concentration should exceed 3.6% (w/v).

**Table 9.1** Base materials for the production of vinegar.

| | |
|---|---|
| Apple | Palm sap |
| Banana (and skins) | Peach |
| Cashew apples | Pear |
| Cocoa sweatings | Persimmon |
| Coconut water | Pineapple |
| Coffee pulp | Prickly pear |
| Dates | Prune |
| Ethanol | Rice |
| Honey | Sugar cane |
| Jackfruit | Sweet potato |
| Jamun | Tamarind |
| Kiwi fruit | Tea |
| Malted barley | Tomato |
| Mango | Watermelon |
| Maple products | Whey |
| Molasses | Wine |
| Orange | |

The key organism is Acetobacter (formerly known as Mycoderma), with pertinent strains being *Acetobacter aceti*, *Acetobacter pastorianus* and *Acetobacter hansenii*. Depending on the species, they function best in the temperature range 18–34°C. Fermentation is usually arrested when there is a minimal but finite residual ethanol presence so as to avoid over-oxidation to $CO_2$ and water. The key equation is

$$CH_3CH_2OH + O_2 \rightarrow CH_3CO_2H + H_2O$$

The conversion of ethanol to acetic acid is accompanied by secondary fermentation important for the generation of aroma-active compounds, such as acetaldehyde, ethyl acetate and other esters, and higher alcohols, such as methyl butanol. The flavour so-derived (and also directly) depends on the source of the alcohol.

## Vinegar making processes

The slow Orleans process is employed for the manufacture of high-quality vinegars (Fig. 9.1). The starting liquor is held in large casks containing wood shavings or grape stalks that represent a large surface area on which the microbes can thrive. Acetification commences and after 8 days, the liquid is withdrawn and transferred to barrels so as to become half to two-thirds full. Fresh vinegar stock is introduced into the main cask to replace that which has been removed. Acidity reaches a maximum after approximately 3 months. On a weekly basis, one-quarter to two-thirds of the contents are removed from the base of each barrel to be replaced from the main cask.

Other processes aim at closer contact of liquid and organism, presenting the highest possible surface area so as to facilitate access of oxygen, thereby reducing the time for acetification. Tanks of wood or steel incorporate cooling coils (temperature maintained at 27–30°C) and are vented to allow circulation of air. They feature false bottoms to support wood shavings (preferably beech) or grape stalks. There is a spray mechanism to further facilitate rousing (Fig. 9.2) and distribution. The liquid trickles over the support and is pumped back to a header tank. Acetification will be complete after approximately

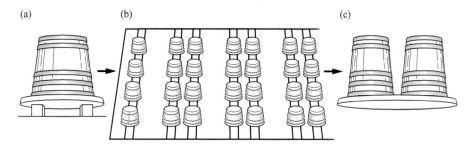

**Fig. 9.1** The Orleans process. (a) Starting vat, (b) vats for acetification and (c) vats for clarifying.

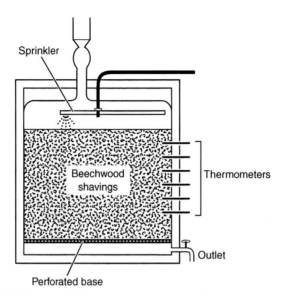

Fig. 9.2   A vinegar generator.

1 week. A proportion of the vinegar is removed from the base of the tanks and replaced with an equal volume of fresh feedstock. Some 20% evaporative loss occurs and the shavings must be replaced annually.

The submerged process, which is now the main approach, does not employ wood shavings and depends on carefully selected cultures of Acetobacter growing in aerated deep culture. It is conducted in tanks of stainless steel or polypropylene reinforced with fibreglass and with capacities of up to 120 hL. The vessel incorporates systems to ensure continuous flow of air and also coils to maintain a temperature of around 30°C. Oxidation starts slowly and air is introduced hourly to permeate completely. Acetification is complete when 0.2–1.5% (w/v) alcohol survives. It is a very rapid process. About half of the vinegar is bled off, with the remainder acting as the 'mother' for the next batch. Yields are high (90–95%) due to much less loss by evaporation than in the other approaches. However, the vinegar tends to be more cloudy and less aromatic, as there is less opportunity for flavour development to occur, for example that catalysed by the esterases.

Finally the vinegar is filtered and perhaps loaded into wooden casks to allow ageing. Vinegar is customarily matured in sealed, completely filled vats of stainless steel or wood for up to 1 year to allow flavour refinement and settling of insolubles. Bentonite is the most common clarification agent employed.

## Malt vinegar

Malting of barley and ensuing mashing and fermentation are exactly analogous to the approaches for beer (see Chapter 2). However, of course, no hops are used in the boiling stage. Adjuncts such as corn or rice may be used. The alcoholic solution obtained is separated from the yeast and inoculated with Acetobacter. Such vinegar must contain at least 4% w/v acetic acid.

Distilled malt vinegar (colourless) is made by the distillation of malt vinegar and is used, for example, in the pickling of onions.

## Wine vinegar

This is the main vinegar on the continent of Europe, and is made from low alcohol wines (7–9%) or from those with too high volatile acidity. Any wines that have too high an alcohol content must be diluted; otherwise, the Acetobacter will be inhibited. Too high a sulphur dioxide level or sediment level will also be a problem. When produced on a small scale, the wine is mixed in small wooden barrels with mother vinegar. The barrel must contain air so it is not filled completely. The process halts naturally when the acetic acid content reaches 7–8% w/v. The product will contain elevated levels of acetaldehyde and ethyl acetate when compared with the parent wine. Some of the vinegar will now be drawn off for use and replaced with fresh wine. Production on a larger scale is subject to EU regulations, with the stipulation that the total acid developed must be greater than 6% w/v and the maximum surviving ethanol being less than 1.5% v/v.

## Other vinegars

Cider vinegar is produced from hard cider or apple wine, has a yellow hue and may be coloured further with caramel. Such ciders tend to have a relatively low acidity. Vinegars may be made from a range of other fermented fruits, taking on some of the character of the original base.

Rice vinegar derives from the acetification of sake or its co-products. When compared with cider vinegar, rice vinegar tends to have a fairly low acidity and has a light and delicate flavour highly favoured for oriental cooking because of its low impact on the flavour imparted by the other materials in the dish.

Molasses has been used as a base for vinegar production (though not extensively) as a mechanism for dealing with by-products of the sugar industry. Mead has been employed as a vinegar base, too.

Spirit vinegar, sometimes called white distilled vinegar, is derived from alcohol obtained by the distillations of fermented sugar solutions. If legally permitted, synthetic ethanol is used, diluted to 10–14% ABV. It is colourless of course, but may be darkened by the addition of caramel. As is to be expected, this is the cheapest vinegar to produce and, accordingly, is the one that is most widespread for general use and, when diluted to 4–5%, for use in pickling.

## Chemical synthesis of vinegar

Acetic acid can be produced by the catalytic oxidation of acetaldehyde, which in turn is produced by the catalytic hydration of acetylene or by the catalytic dehydrogenation of ethanol. The undesirable formic acid and formaldehyde are eliminated by distillation. The acetic acid is purified before diluting to 60–80% by volume to obtain the vinegar essence. This in turn is diluted to 4–5% in the generation of food grade 'vinegar'. Sugar, salt and colour may be added. In the United Kingdom, such a product must be labelled 'non-brewed condiment'.

## Balsamic

At the other end of the quality spectrum is balsamic vinegar. It has been produced for hundreds of years in Northern Italy, notably the provinces of Modina and Reggio Emilia. The base material is grape must, preferably Trebbiano. Alcoholic fermentation is effected about 24 h after pressing, with must gently boiled until it is reduced to a third or a half by volume. This leads to a high sugar concentration of about 30%. The alcoholic fermentation and the acetification occur together very slowly. The relevant organisms are yeasts Saccharomyces and Zygosaccharomyces and bacteria Acetobacter and Gluconobacter. In the process, a series of chemical transformations alongside the slow microbial action leads to a flavoursome and complex mix of alcohols, aldehydes and organic acids.

The process is performed in a series of decreasingly sized barrels made of various types of wood. They are located in efficiently ventilated areas that are hot and dry in the summer months but cool in winter. Each year a portion from the smallest barrel is removed for consumption to be replaced by an equivalent amount from the next sized barrel, which in turn has its volume restored from the next barrel, and so on. The largest barrel is made up to volume using that season's boiled must. The finished product is dark brown, syrupy, sweet, sour (6–18% acetic acid by weight) and with a pleasant aroma. This patient process takes at least a dozen years, with some products emerging for sale after as many as 50 years. Yields are perforce low (less than 1 L of vinegar from 100 kg of fresh must).

The chemical composition and major volatile components of the main vinegars are shown in Tables 9.2 and 9.3, respectively.

**Table 9.2**  Chemical composition of vinegars.

| Parameter | Balsamic | Cider | Malt | Wine | Synthetic |
|---|---|---|---|---|---|
| Specific gravity | 1.042–1.361 | 1.013–1.024 | 1.013–1.022 | 1.013–1.02 | 1.007–1.022 |
| Total solids $(g L^{-1})$ | 337–874 | 19–35 | 3.0–28.4 | 8.7–24.9 | 1.0–4.5 |
| Total acidity (as acetic acid, %) | 6.2–14.9 | 3.9–9.0 | 4.3–5.9 | 5.9–9.2 | 4.1–5.3 |
| Sugars $(g L^{-1})$ | 351–690 | 1.5–7.0 | — | 0–6.2 | — |

Date derived from Plessi (2003).

**Table 9.3** Volatile components in vinegars.

| Volatile | Balsamic | Cider | Malt | Wine |
|---|---|---|---|---|
| Acetaldehyde | ✓ | ✓ | ✓ | ✓ |
| Acetone | ✓ | | ✓ | ✓ |
| Benzaldehyde | | ✓ | ✓ | |
| 2,3-Butanediol | ✓ | | | ✓ |
| 2,3-Butanedione | ✓ | | | |
| 2-Butanone | ✓ | | | |
| γ–Butyrolactone | | | | ✓ |
| Diethyl succinate | | ✓ | | |
| Ethanol | ✓ | ✓ | ✓ | ✓ |
| Ethyl acetate | ✓ | ✓ | ✓ | ✓ |
| Ethyl formate | ✓ | ✓ | ✓ | ✓ |
| Ethyl lactate | | ✓ | | |
| Furan | ✓ | | | |
| Furfural | ✓ | | | |
| 3-Hydroxy-2-butanone | ✓ | | | ✓ |
| Isobutanal | ✓ | | | |
| Isobutyl acetate | | | ✓ | ✓ |
| Isobutyl formate | | | ✓ | ✓ |
| Isopentyl acetate | | | ✓ | ✓ |
| Isopentyl formate | | | ✓ | |
| Isovaleraldehyde | | | | ✓ |
| Methyl acetate | ✓ | | | |
| 2-Methylbutanal | ✓ | | | |
| 2-Methyl-1-butanol | | ✓ | ✓ | ✓ |
| 3-Methyl-1-butanol | ✓ | ✓ | ✓ | ✓ |
| 2-Methyl-1-propanol | | ✓ | ✓ | ✓ |
| 2-Methyl-3-butene-2-ol | | ✓ | | |
| 2-Pentanone | | ✓ | ✓ | ✓ |
| 2-Pentanol | | ✓ | | |
| 3-Pentanol | | ✓ | | |
| Phenylacetaldehye | | ✓ | | |
| Propionaldehyde | | ✓ | | |
| 2,4,5-Trimethyl-1,3-dioxolane | ✓ | | | |

# Bibliography

Conner, H.A. & Allgeier, R.J. (1976) Vinegar: its history and development. *Advances in Applied Microbiology*, **20**, 81–133.

Plessi, M. (2003) Vinegar. In *Encyclopedia of Food Sciences and Nutrition* (eds B. Caballero, L.C. Trugo & P.M. Finglas), pp. 5996–6003. Oxford: Academic Press.

Plessi, M. & Coppini, D. (1984) L'Aceto balsamico tradizionale di Modena. *Atti della Società dei Naturalisti e Matematica*, **115**, 39–46.

# Chapter 10
# **Cheese**

Cheese making can be traced back some 8000–9000 years to origins in the Fertile Crescent, that is, latter day Iraq. Just as beer arose from the adventitious contamination of moist sprouted grain, so did cheese develop as a consequence of the accidental souring of milk by lactic acid bacteria, with the attendant clotting to produce curd. Cheese, whey (the liquid that separates from the curd) and fermented milks all comprise milk rendered as long life forms. The first enzyme employed to curdle milk was obtained unknowingly (the first cell-free enzyme preparation not having been made until 1897, by Buchner from brewer's yeast) from the stomachs of the hare and kid goats that were immersed in milk. Rennin was not produced in an isolated form from calf vells until 1970. Similarly, adventitious organisms are less widely used for cheeses nowadays – and pure cultures of lactic acid bacteria have been available since 1890.

Parallels between cheese making and the production of beer (and many other fermented foods) continue when one considers the evolution of the modern cheese making business. The Industrial Revolution with the advent of extensive rail networks and heavy, urbanisation to support expanding employment in large factories meant that cheese production was consolidated into a relatively few large producers employing enhanced control and automation.

There are in excess of 2000 different types of cheese. The Food and Agriculture Organisation (FAO) definition of cheese is

> Cheese is the fresh or matured product obtained by the drainage (of liquid) after the coagulation of milk, cream, skimmed or partly skimmed milk, butter milk or a combination thereof. Whey cheese is the product obtained by concentration or coagulation of whey with or without the addition of milk or milk fat.

One can classify cheeses according to their country of origin, composition, firmness and which maturation agents are employed in their production and by the processes generally employed in their manufacture and maturation (Table 10.1). The listing shown does not include the spiced cheeses that incorporate the likes of caraway seeds, cloves, cumin and peppers.

An overview of cheese making is given in Fig. 10.1. The critical requirement is that the cheese should have the correct pH and moisture content. Easily the most important need is to time and control acid production, alongside the control of expulsion of the whey that contains the substrates and buffers that regulate how much acid is produced and the extent to which pH changes occur.

Unless cheese is heat-processed, its composition will continually change through the action of surviving micro-organisms and enzymes.

**Table 10.1**  Some types of cheese.

| Firmness and subdivision | Moisture | Examples |
|---|---|---|
| Soft | 50–80% | |
| Unripened/low fat | | Cottage |
| Unripened/high fat | | Cream |
| Unripened stretched curd | | Mozzarella |
| Ripened through external | | Brie |
| mould growth | | Camembert |
| Ripened by bacterial | | Kochkäse |
| fermentation | | |
| Salt-cured or pickled | | Feta |
| Surface-ripened | | Liederkranz |
| Semi-soft | 39–50% | |
| Ripened through internal | | Blue |
| mould growth | | Gorgonzola |
| Surface-ripened by bacteria | | Limburger |
| and yeast | | |
| Chiefly ripened through | | Bel Paese |
| internal bacterial | | Munster |
| fermentation but perhaps | | |
| also surface growth | | |
| Ripened internally by | | Provolone |
| bacterial fermentation | | |
| Hard | <39% | |
| Ripened internally by | | Cheddar |
| bacterial fermentation | | |
| Ripened internally by | | Edam |
| bacterial fermentation, also | | Emmental (Swiss) |
| with 'eye' production | | Gouda Gruyere |
| Ripened by internal mould growth | | Stilton |
| Very hard cheese | <34% | Parmesan |
| Whey cheese | 60% | |
| By heat/acid denaturation | | Ricotta |
| of whey protein | | |

Derived from Olson (1995).

# Milk

The composition of milk is summarised in Table 10.2. For the most part, the milk employed in the production of cheese is from the cow, but essentially any milk can be converted into cheese. The key criteria are the content of protein and of fat.

The proteins, especially the caseins, form the main structural 'architecture' for the cheese. The fat, which comprises spherical globules in the milk, becomes trapped within the protein matrix in the cheese. Carbohydrate, of which lactose is the most important, is for the most part expelled with the whey, the remainder being fermented to lactic acid. The fourth major component is calcium phosphate, much of it in a micellar form, which makes a key contribution to the physical properties of cheese.

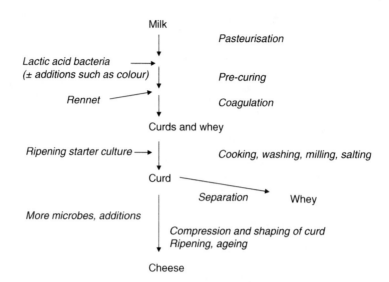

**Fig. 10.1**   Making cheese.

**Table 10.2**   Composition of cow's milk.

| Component | Percentage |
| --- | --- |
| Water | 87.3 |
| Lactose | 4.8 |
| Fat | 3.7 |
| Caseins | 2.8 |
| Whey protein | 0.6 |
| Ash | 0.7 |

The main constituents of the protein fraction are caseins and the whey proteins, the latter being water soluble and therefore expelled with the whey. The caseins are phosphoproteins that precipitate at 20°C from raw milk at pH 4.6. There are three major casein fractions: $\alpha$, $\beta$ and $\kappa$ and they tend to associate via electrostatic and hydrophobic interactions to afford micelles, rendering a colloidal suspension in the milk, one which is impacted by calcium phosphate (Fig. 10.2).

Ninety-six per cent of the lipid is in the form of globules in colloidal suspension. They are coated by emulsion-stabilising membranes in a lipid bilayer with protein at interfaces. This ensures integrity of the globules which, if degraded, release free fats that give an oily mouthful and an undesirable appearance.

Short-chain fatty acids (principally C4:0 and C6:0) contribute to the flavour in certain cheeses. The complexity of flavour in goat and sheep cheese is dependent on these and other fatty acids.

Cow's milk comprises 4.8% lactose. This is either fermented as is or after hydrolysis to glucose and galactose. If it is not efficiently eliminated with the

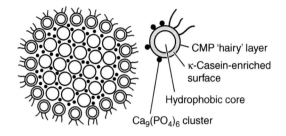

**Fig. 10.2** Micelles in cheese curd (modified from http://www.foodsci.uoguelph.ca/deicon/casein.html).

whey, then it will lead to the risk of colour pick-up in the Maillard reaction and to the growth of spoilage organisms.

Milk also contains some enzymes. Advantage is taken of its heat-sensitive alkaline phosphatase to test for the efficiency of pasteurisation: if the enzyme is destroyed, then this is indicative of sufficient heat having been applied.

The milk may be pretreated in various ways depending on the cheese that is being made. Such treatments may include

(1) heating (pasteurisation) to destroy pathogens and lower the levels of spoilage bacteria and enzymes. Such treatment may typically be a regime of 72°C for 15 s;

(2) reduction of fat by centrifugation or by adding non-fat solids such as concentrated skimmed milk or non-fat dry milk. However, this may be problematic if lactose levels are too high;

(3) concentration, which may be by applying vacuum (for high throughput cheeses) or ultrafiltration (for soft cheeses);

(4) clarification, either by high-speed centrifugation or microfiltration. This procedure optimises the number of foci that lead to 'eyes' in the finished cheese. Very high-speed centrifugation will additionally lower the level of undesirable micro-organisms;

(5) homogenisation. This involves the application of high-pressure shear to disrupt fat globules, rendering smaller globules that are coated with protein. This is important for rendering consistent texture in blue-veined cheeses and for cream cheese. It also has significance for the levels of free fatty acids and therefore of the flavour-active oxidation products that are made from them;

(6) addition of calcium chloride, which promotes clotting;

(7) addition of enzymes to enhance flavour or to accelerate maturation. For example, lipases may be employed in the manufacture of blue-veined cheeses;

(8) addition of micro-organisms. These microbes may include Propionibacter for Emmental and Swiss cheese, *Penicillium roqueforti* for blue cheeses and *P. camamberti* for camembert and brie.

**Table 10.3**  Lactic acid bacteria used in cheese production.

| Cheese type | Organisms |
|---|---|
| Italian grana and pasta types, Swiss | Thermophilics<br>*Lactobacillus delbrueckii* ssp. *bulgaricus*<br>*Lactobacillus helveticus*<br>*Streptococcus thermophilus* |
| Blue, Cheddar, cottage, cream, Gouda, Limburger | Homofermentative<br>*Lactoccus lactis* ssp. *cremoris*<br>*Lactococcus lactis* ssp. *lactis*<br>*Lactococcus lactis* ssp. *lactis* biovar *diacetylactis*[a] |
| Blue, cottage, cream, Gouda | Heterofermentative<br>*Leuconostoc mesenteroides* ssp. *cremoris* |

[a] This organism has a plasmid coding for enzymes that allow the metabolism of citrate.
Based on Olsen (1995).

## The culturing of milk with lactic acid bacteria

Lactic acid bacteria are used in the manufacture of all cheeses except those in which curdling is effected by the application of acidification with or without heating. The classification of bacteria is given in Table 10.3. Important characteristics of the individual strains include their ability to generate lactic acid at various temperatures and their capability for producing carbon dioxide and diacetyl that are important for the appearance (e.g. 'eyes' in Gouda) and flavour (e.g. in Cottage cheese). Diacetyl may also serve a valuable role as an antimicrobial agent, as might also organic acids and hydrogen peroxide generated by lactic acid bacteria. The natural antimicrobial nisin is permitted for use in some countries, but a more common preservative is potassium sorbate.

Various sizes and shapes of vats are employed. Commodity cheeses such as Cheddar will tend to be produced in very large mechanised vessels. Speciality chesses however will emerge from small, less extensively mechanised vats.

The rate of addition of lactic acid bacteria must be carefully regulated not only for efficiency in the process but also to ensure consistency in the product. Modern cheese making facilities will incorporate sophisticated propagation and inoculation control regimes. It is increasingly the case that the organisms are supplied as starter cultures from commercial suppliers.

## Milk clotting

The gel must be uniform and possess the appropriate strength in order that there should be maximum retention of casein and milk fat, as well as to minimise variation in the levels of moisture. Enzymes are preserved by ensuring that the temperature does not rise excessively and protecting the process stream from excesses of pH and oxidising agents such as the hypochlorites employed in cleaning regimes.

The most important milk clotting enzyme is chymosin, which has an optimum pH around 6.0. A shortage of calves, alongside public acceptability issues, mean that alternative source of the enzyme have been sought. The gene for chymosin has been expressed in microbes, notably *A. niger*, *K. lactis*, and *E. coli* K12. More than half of the world cheese market is probably now dependent on the use of such preparations.

Clotting occurs due to the hydrolysis of a single bond in $\kappa$-casein, the impact of which is reduced micelle stabilising capability. The hydrolysis releases the hydrophilic N-terminal region of the molecule which in the unhydrolysed molecule serves the function of reaching out from the micelle surface into the solvent and stabilising it. Accordingly, the micelles aggregate. Enzyme activity is also important for the initial proteolysis during cheese maturation.

Cheeses differ in their optimum gel firmness. Those that have firmer gels will expel whey more slowly.

## Whey expulsion

Whey is expelled rapidly from the curd after it has been cut into small pieces. This will be further accelerated by an increase in temperature when the mix is agitated.

Lactic acid bacteria trapped in the curd metabolise lactose to lactic acid and this diffuses from the curd. The rate at which this occurs, as well as the rate at which moisture and lactose are removed, have substantial impact on the nature of the finished cheese.

Whey expulsion also has an impact on the release of calcium phosphate from the casein matrix. Calcium phosphate greatly influences the physical properties of casein aggregates and the more it is removed, the more brittle the cheese is. The calcium phosphate-casein structure is also influence by pH, which in turn depends on the extent of lactic acid production and the buffering capacity of the curd. The buffering capacity depends on the concentration of undissociated calcium phosphate, casein and lactate surviving in the cheese. pH also influences the action of the milk clotting enzymes, a lower pH allowing better survival of chymosin and, in turn, a more brittle cheese.

## Curd handling

The curd is separated from whey by settling and drainage through some form of perforated system. It is important to have efficient fusion of the curd particles and this is impacted by pH and by the physical properties of the curd. Fusion starts to occur when the pH has reached 5.8, and if the whey is removed before this, the cheese will feature openings. If fusion takes place in the presence of whey, the cheese will have a dense body. Sodium chloride may be introduced into the curd after the whey has been drained, a process that controls acid production and impacts the final flavour.

Finally the curd particles are fused into the desired final shape. This may be promoted by increased pressure or by the application of vacuum. Fused cheeses usually receive protection from moulds and other microbes by coating with wax or application of a plastic film. However, cheeses such as Camembert are not sealed immediately in order that there is an opportunity for microbial growth.

## The production of processed cheese

Processed (or process) cheese is made by heating and mixing combinations of cheese and other ingredients, with the end result being a creamy, smooth product of desirable texture, flavour, appearance, and physical attributes, such as melting and flow properties. Processed cheese incorporates phosphates and citrates that prevent the separation of oil and protein phases during heating. The phosphates and citrates bind minerals in the cheese increasing the solubility of caseins. The proteins form a thin film around the fats which are thus stabilised against separation.

## The maturation of cheese

Most cheeses are matured for periods between 3 weeks and more than 2 years, the period being essentially inversely proportional to the moisture content of the cheese. This comprises controlled storage to allow the action of enzymes and microbes to effect desired physical and flavour changes. Amongst the changes that occur are the bacterial reduction of lactose to lactate (via glycolysis) in eye cheeses, mould-ripened cheeses and smear-ripened cheeses, and the conversion of citrate *inter alia* to acetate, diacetyl and acetoin.

Proteolytic cleavage of $\alpha$-casein is important for the softening of cheeses such as Gouda and Cheddar. Furthermore, amino acid production by

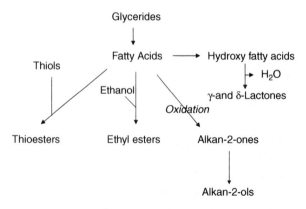

**Fig. 10.3**   Reactions of lipids involved in the development of cheese flavour.

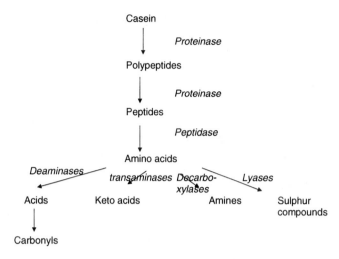

**Fig. 10.4**  Reactions of proteins involved in the development of cheese flavour.

**Table 10.4**  Examples of flavour-active volatiles in cheese and their origins.

| Substance | Origin | Route |
|---|---|---|
| 3-Methylpropionic acid | Leucine | Deamination |
| 2-Keto-4-methylpentanoic acid | Leucine | Transamination |
| 3-Methylbutanal | Leucine | Decarboxylation of 2-keto-4-methylpentanoic acid |
| 4-Methylpentanoic acid | Leucine | Deamination |
| 3-Methylbutanal | Leucine | Reduction of 3-methylbutanal |
| 3-Methylbutanoic acid | Leucine | Oxidation of 3-methylbutanal |
| 4-Methylthio-2-ketobutyrate | Methionine | Transamination |
| Methional | Methionine | Decarboxylation of 4-methylthio-2-ketobutyrate |
| 2-Keto-4-thiomethylbutyrate | Methionine | Deamination or transamination |
| Methanethiol | Methionine | Demethiolation of 2-keto-4-thiomethylbutyrate |
| Dimethyl disulphide | Methionine | Degradation of methanethiol |
| Dimethyl sulphide | Methionine | Degradation of methanethiol |
| Hydrogen sulphide | Methionine | Degradation of methanethiol |
| Dimethyl trisulphide | Methionine | Addition of sulphur to dimethyl disulphide |
| Methyl thioacetate | Methionine | Reaction of methanethiol with acetyl-CoA |
| Tyramine | Tyrosine | Decarboxylation |
| *p*-Hydroxyphenylpyruvate | Tyrosine | Transamination |
| *p*-Cresol | Tyrosine | Via *p*-Hydroxyphenylpyruvate |
| Indolepyruvate | Tryptophan | Transamination |
| Skatole | Tryptophan | Via Indolepyruvate |
| 2-Methyl butanal | Isoleucine | Strecker degradation |

proteolytic enzymes, including aminopeptidases, may be important for the growth of organisms that function in maturation. Figures 10.3 and 10.4 illustrate some of the biochemical pathways that can occur during maturation of cheeses. Table 10.4 lists a range of flavour-active substances in cheese derived from these and other reactions. This is a very restrictive list and is meant to be merely illustrative of the way in which the wide diversity of flavour-active species arise.

## Bibliography

Fox, P.F., ed. (1993) *Cheese: Chemistry, Physics and Microbiology – General Aspects*. London: Chapman & Hall.

Fox, P.F., Guinee, T.P., Cogan, T.M. & McSweeney, P.L. (2000) *Fundamentals of Cheese Science*. Gaithersburg, MD: Aspen.

Law, B.A., ed. (1997) *Microbiology and Biochemistry of Cheese and Fermented Milk*. London: Blackie.

Law, B.A., ed. (1999) *Technology of Cheesemaking*. Sheffield: Sheffield Academic Press.

Olson, N.F. (1995) Cheese. In *Biotechnology*, 2nd edn, vol. 9, Enzymes, Biomass, Food and Feed (eds H.-J. Rehm & G. Reed), pp. 353–384. Weinheim: VCH.

Robinson, R.K. & Wibey, R.A. (1998) *Cheesemaking Practice*. Gaithersburg, MD: Aspen.

Scott, R. (1986) *Cheesemaking Practice*, 2nd edn. London: Elsevier.

Wong, N.P., ed. (1988) *Fundamentals of Dairy Chemistry*, 3rd edn. New York: Van Nostrand Reinhold.

# Chapter 11
# Yoghurt and Other Fermented Milk Products

Like cheese, yoghurt originated as a vehicle to preserve the nutrient value of milk. Through time, the product has evolved to a foodstuff richly diverse in flavour, texture and functional properties. Thus, the formulations may now incorporate components such as fruits, grains and nuts, as well as having a range of textures.

Yoghurt is only one of a series of fermented dairy products (Table 11.1). Sour cream comprises cream (>18% milk fat) fermented with specific lactic cultures, perhaps with the use of rennin, flavours and materials to enhance texture. Kefir and kourmiss are fermented milks from Russia and Eastern Europe. In their production, yeast accompanies bacteria with the impact of producing alcohol and carbon dioxide. Rather than seeding with organisms, an endogenous microflora is employed and this supposedly contributes to the health value of such products. The organisms employed are listed in Table 11.2.

Basically yoghurt is a semisolid foodstuff made from heat-treated stabilised milk through the action of a 3:1 mixture of *Streptococcus salivarus* ssp. *thermophilus* (ST) and *Lactobacillus delbrueckii* ssp. *bulgaricus* (LB). Their relationship is symbiotic. In some countries, other organisms are also used, namely *L. acidophilus* and *Bifidobacterium* spp.

Starter cultures are purchased either in a freeze-dried, liquid nitrogen or frozen form. They are used either as is or receive further propagation. This is in liquid skim milk or a blend of non-fat dry milk in water (9–12% solids). Media may also include citrate, which is a precursor of the diacetyl that makes a major contribution to flavour.

The milk used originates from a range of animals, but is chiefly from the cow. To achieve the desired consistency, the milk is fortified with dried or condensed milk. Vitamin A (2000 IU per quart) and vitamin D (400 IU per quart) may also be added. Other additions sometimes used are lactose or whey to increase the content of non-fat solids; sucrose, fructose or maltose as sweeteners; flavourings, colour, and stabilisers.

Milk is the natural habitat for a range of lactic acid bacteria. Milk of course will spontaneously sour, but the uncontrolled nature of this means that starter cultures are nowadays the norm.

**Table 11.1**   Examples of fermented dairy foods other than cheese.

| Foodstuff | Description | Origin |
|---|---|---|
| Acidophilus milk | Low-fat milk. Heat-treated and inoculated with *Lactobacillus acidophilus* or *Bifidobacterium bifidum* | USA, Russia |
| Chal | Camel's milk yoghurt | Turkmenistan |
| Cultured buttermilk | Skim cow's milk heated, homogenised, cooled and inoculated with *Streptococcus cremoris, Streptococcus lactis, Streptococcus lactis* ssp. *diacetylactis, Leuconostoc cremoris* | USA |
| Filmjolk | Whole cow's milk pasteurised, homogenised, cooled, fermented with ropy strains of *Streptococcus cremoris* and other organisms used for cultured buttermilk. The polymers giving ropiness are important for the slimy texture | Sweden |
| Kefir | Acidic and mildly alcoholic effervescent milk. Goat, buffalo or cow milk heated to 90–95°C for 3–5 min, cooled and inoculated in an earthenware vessel with Kefir grains or starter comprising *Lactobacillus casei, Streptococcus lactis, Lactobacillus bulgaricus, Leuconostoc cremoris, Candida kefyr, Kluyveromyces fragilis*, etc. | Russia |
| Kumiss | Similar to Kefir, from horse milk and frequently served with cereal | Russia |
| Lassi | Sour drink consumed salted with herbs and spices or sweetened with honey. | India |
| Quark | Low-fat acidic soft cheese eaten fresh. Fresh milk pasteurised, cooled, treated with rennet and starter culture of lactic acid bacteria (similar population to cultured buttermilk) | Germany |
| Ricotta | Hard cheese from whey, used as whipped dessert or for making of gnocchi or lasagne. Whey, perhaps with added skimmed, whole milk or cream, salt and *Streptococcus thermophilus* and *Lactobacillus bulgaricus*, followed by heat treatment and curd collection | Europe |

As raw milk contains heat-sensitive microbial inhibitors, notably the enzyme lysozyme and agglutinins, it is either heated at 72°C for 16 s or autoclaved for 15 min at the onset of the process. This heating also degrades casein, liberating thiol groups and it also encourages the shift of lactose to lactic acid.

The non-fat solid content of milk varies seasonally and this in turn impacts the microflora, with greater growth of lactic acid bacteria as the solid content increases.

The bacteria are also at risk of bacteriophage infection, for which reason chlorine (200–300 ppm) is applied to processing equipment, and culture rooms are fogged with 500–1000 ppm chlorine. Culture media may also incorporate phosphate to sequester the calcium that is needed for phage growth.

The production of lactic acid must be sufficient to lower the pH to a level where acetaldehyde and diacetyl (amongst other flavour-active components) are generated sufficiently.

**Table 11.2**  Organisms involved in making fermented milks.

| Foodstuff | Organisms |
| --- | --- |
| Acidophilus milk | *Lactobacillus acidophilus* |
| Cultured buttermilk | *Lactoccus lactis* ssp. *cremoris*<br>*Lactococcus lactis* ssp. *lactis*<br>*Lactococcus lactis* ssp. *lactis* biovar *diacetylactis* |
| Kefir | *Lactoccus lactis* ssp. *cremoris*<br>*Lactococcus lactis* ssp. *lactis*<br>*Lactobacillus delbrueckii* ssp. *bulgaricus*<br>*Lactobacillus helveticus*<br>*Lactobacillus delbrueckii* ssp. *lactis*<br>*Lactobacillus casei*<br>*Lactobacillus brevis*<br>*Lactobacillus kefir*<br>*Leuconostoc mesenteroides*<br>*Leuconostoc dextranicum*<br>*Acetobacter aceti*<br>*Candida kefir*<br>*Kluyveromyces marxianus* ssp. *marxianus*<br>*Saccharomyces cerevisiae*<br>*Torulospora delbrueckii* |
| Kumiss | *Lactobacillus delbrueckii* ssp. *bulgaricus*<br>*Lactobacillus kefir*<br>*Lactobacillus lactis*<br>*Acetobacter aceti*<br>*Mycoderma* sp.<br>*Saccharomyces cartilaginosus*<br>*Saccharomyces lactis* |
| Yoghurt | *Lactobacillus delbrueckii* ssp. *bulgaricus*<br>*Streptococcus salivarius* ssp. *thermophilus* |

# Bibliography

Kosikowski, F.V. (1982) *Cheese and Fermented Milk Foods*, 2nd edn. Brooktondale: Kosikowski.

Robinson, R.K., ed. (1986) *Modern Dairy Technology, volume II. Advances in Milk Products*. London: Elsevier.

Robinson, R.K., ed. (1992) *Therapeutic Properties of Fermented Milks*. New York: Elsevier.

Tamime, A.Y. & Robinson, R.K. (1999) *Yoghurt: Science and Technology*. Cambridge: Woodhead.

Wood, J.B., ed. (1992) *The Lactic Acid Bacteria*. London: Elsevier.

# Chapter 12
# **Bread**

Despite the seeming ludicrousness of certain well-publicised latter-day low carbohydrate diets, bread remains a staple food for numerous people worldwide, representing perhaps as much as 80% of the dietary intake in some societies.

Like beer, its origins can be traced to the gruel obtained from mixing ground grain (notably barley in the earliest times) with water or milk. The blend was then subjected to air-drying or was baked either on hot stones or by being put into hot ashes, such ovens being traced to early Babylonian civilisation.

Such breads broken into water and allowed to spontaneously ferment in jars were of course the origins of beer. Preferences for bread *per se* shifted from a flat form to loaves, and wheat replaced barley as the main raw material, although rye has long played a major role in bread making in central and northern Europe.

Without of course knowing the science involved, the Egyptians were producing leavened bread and soured dough can be traced to 450 BC.

In more modern times, the first dough kneading machines were developed late in the eighteenth century, while large-scale commercial production of baker's yeast commenced in the nineteenth century. And as for other fermentation products described in this book, it was the Industrial Revolution that led to the emergence of large commercial bakeries. Breads assumed much more uniformity in quality, size and shape. However, the local variation still prevalent in terms of styles of bread, whether loaves or flat breads, is at least the equal of variation in most other products of fermentation.

Bread made from flour and water but no leavening agent is flat, for example, tortilla, nan. Other breads are leavened by gases or by steam, this demanding that the doughs are capable of holding gas.

The key ingredients in the production of bread are grain starch (chiefly wheat or rye), water, salt and a leavening agent. Sometimes sugar, fat and eggs are amongst the additional components, while acids are used in the production of rye breads. Whereas wheat doughs are leavened with yeast, rye doughs are not only treated with yeast but also acidified by sourdough starter cultures or acid *per se*. Gas retention in wheat doughs is dependent upon the gluten structure, whereas in rye doughs there is less retention of gas and the presence of mucilage and a high dough viscosity is important.

An overview of bread production is given in Fig. 12.1. The key steps are (1) preparation of raw materials; (2) dough fermentation and kneading;

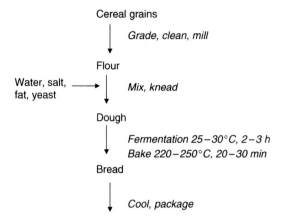

**Fig. 12.1** Making bread.

(3) processing of the dough (fermentation, leavening, dividing, moulding and shaping); (4) baking; (5) final treatments, such as slicing and packaging.

## Flour

The major functional component within wheat flour is its protein, gluten. The gluten must have good water absorbing properties, elasticity and extensibility. The cereal starch should be readily gelatinised because the production of maltose is important if the yeast is to be able to 'raise' the bread. The precise significance of the gluten varies between bread types. For instance, crackers demand low protein content and weak gluten. Chemically leavened products such as cookies require flours that afford 'shortness': the gluten concentration is low but the starch has good pasting characteristics. By contrast, the baking quality of rye flours is very much determined by the properties of the pentosans and starch.

## Water

The ionic composition of the water is important, and the hardness is preferably in the range 75–150 ppm. Carbonates and sulphates allow firmer and more resilient gluten.

## Salt

Typically there is 1.5–2% salt in most breads. While of primary significance for flavour, sodium chloride also inhibits the hydration of gluten, rendering it shorter. This means that the doughs do not collapse and gas retention is enhanced. If no salt is employed, then there is an increase in dough extension and the dough is moist and runny.

## Fat

Fat makes baked goods shorter by forming a film between the starch and the protein.

## Sugar

Sugar is used to promote fermentation and also browning through the Maillard reaction. It also tends to makes dough more stable, more elastic and shorter.

## Leavening

The main leavening agent is yeast. When yeast was not available, sourdoughs were employed. Their active constituents were in part not only endogenous yeasts but also heterofermentative lactic acid bacteria.

Baker's yeast is of course *Saccharomyces cerevisiae*. It is a top fermenting organism, cultured on molasses in aerobic, fed-batch culture so as to maximise yield. Growth is optimal at 28–32°C and within the pH range 4–5.

The bread mix will comprise 1–6% yeast depending on the weight of flour and some other factors. The yeast is most commonly employed as a compressed cake of 28–32% solids. The cake can be stored at 4°C for 6–8 days and may be mixed with water before use. The yeast may also be in the form of a cream, which is a centrifuged and washed suspension of approximately 18% solids. This is shipped as needed to bakeries for use within the day. For logistical reasons, there is increasing use of ADY, a dehydrated form of 92–96% solids. It can be stored for upwards of a year. It is re-hydrated prior to use.

Sourdough starter cultures typically comprise $2 \times 10^7$ to $9 \times 10^{11}$ per gram bacteria and $1.7 \times 10^5$ to $8 \times 10^6$ per gram yeasts. The precise populations are frequently ill defined, but Lactobacilli are prevalent (Table 12.1). The organisms are either anaerobes or microaerophiles, are either homofermentative or heterofermentative, and are acid tolerant. The acid produced by these organisms results in bread with good grain texture and an elastic crumb. The heterofermentative organisms tend to give preferred organoleptic characters to the product. Thus, San Francisco sourdough employed chiefly the heterofermentative *Lactobacillus sanfranciscensis* and the yeasts *Torulopsis holmii*, *Saccharomyces inusitus* and *Saccharomyces exiguous*.

**Table 12.1**   Sourdough starter organisms.

Homofermentative organisms
   *Lactobacillus acidophilus*
   *Lactobacillus casei*
   *Lactobacillus farciminis*
   *Lactobacillus plantarum*

Heterofermentative organisms
   *Lactobacillus brevis*
   *Lactobacillus brevis* var *lindneri*
   *Lactobacillus buchneri*
   *Lactobacillus fermentum*
   *Lactobacillus fructivorans*

Yeasts
   *Candida crusei*
   *Pichia saitoi*
   *Saccharomyces cerevisiae*
   *Torulopsis holmii*

Chemical leavening agents tend to be employed for sweet goods and cakes. A combination of carbonate and acid when heated generates carbon dioxide. Thus, a mixture of baking powder (sodium bicarbonate) and tartaric acid or citric acid achieves widespread use. Baking powder may also be used to support the leavening power of yeast. Similarly, lactic acid bacteria may accompany baking powder.

Leavening may also be achieved by physical treatments – that is, the beating in of air. Egg whites may be added to underpin foam formation.

One example of mechanical leavening involves the retention of steam between thin sheets of dough and intervening fat layers, namely puff pastry.

## Additives

A range of additional ingredients may be used to gain mastery over variations in raw materials and process conditions. Amongst the enzymes that may be used are pentosanases, which reduce viscosity, notably in rye-based breads, and allow more consistency in water binding. Proteinases afford slacker dough by degrading protein structure. Furthermore, they promote browning and aroma by releasing free amino compounds that enter into Maillard reactions. Emulsifying agents may be used, such as sodium stearoyl lactylate and sorbitan esters (Fig. 12.2). Oxidising agents are used to improve the rheology of the dough such that gas retention is improved. Such agents promote the oxidation of thiol groups in protein to dithiol bridges and the resultant cross-linking of proteins molecules leads to firmer gluten (Fig. 12.3). A key agent is ascorbic acid, which is converted to dehydroascorbic acid during dough preparation and it is the latter that oxidises the thiol groups. Bromate promotes spongy, dry extensible dough with good gas retention. The converse

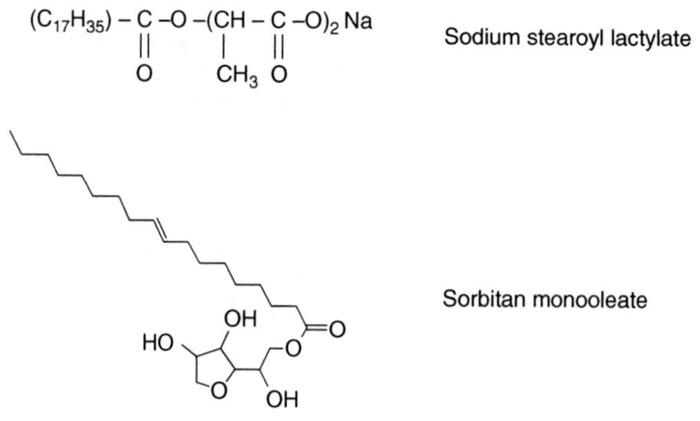

$$(C_{17}H_{35}) - \overset{\displaystyle O}{\underset{\displaystyle \|}{C}} - O - (CH - \overset{\displaystyle O}{\underset{\displaystyle \|}{C}} - O)_2 Na$$
                                    $CH_3$

Sodium stearoyl lactylate

Sorbitan monooleate

**Fig. 12.2** Emulsifying agents.

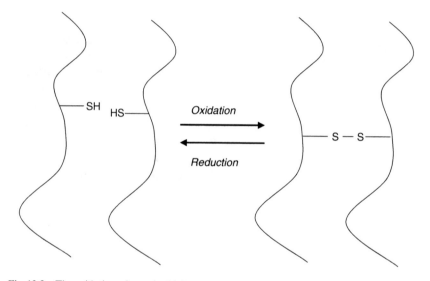

**Fig. 12.3** The oxidation of protein thiol groups.

impacts are afforded by reducing agents (e.g. the couple of cysteine and ascorbic acid), which weaken gluten by breaking thiol bridges. This is important in the making of cookies.

# Fermentation

The yeast requires fermentable sugars, which are produced during the dough phase. Damaged starch is susceptible to the action of endogenous $\alpha$-amylase and $\beta$-amylase and exogenous amyloglucosidase and $\alpha$-amylase. If enzyme levels are insufficient, then loaf volumes and/or flavour are inadequate, the product is crumbly and there is rapid staling. Malt is added as an enzyme source especially for rolls and buns. The resultant increase in sugar causes

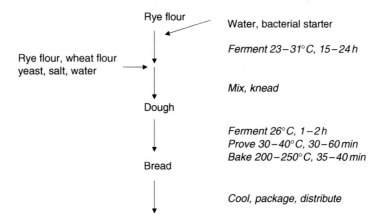

**Fig. 12.4**   The sourdough process.

increased caramelisation and therefore development of colour and flavour and improved crispness and shelf life. The presence of proteolytic enzymes in malt precludes its use in the manufacture of any product demanding strong gluten.

## Dough acidification

This involves the use of either sourdough (Fig. 12.4) or added acids, such as lactic, acetic, citric and tartaric.

## Formation of dough

Dough formation demands good mixing and aeration. The carbon dioxide produced during fermentation increases the size of air bubbles that are introduced, and in turn the oxygen whipped in is utilised by the yeast in its production of membrane materials. The oxygen also has a direct impact on dough structure.

Flour must be stored for 2–4 weeks before it used. The impact is shorter gluten through oxidative events occurring in the storage. Storage must not be prolonged so as to avoid the production of fatty acids that change the rheological properties of the flour and lead to off flavours.

Flour is first sieved, which in itself aids the uptake of air. Mixing with water is performed in diverse types of machine, and must be longer for stronger glutens. Wheat bread dough is mixed at 22–24°C, rye dough at 28°C.

The water hydrates the flour particles with starch absorbing up to a third of its weight. The pentosans also bind water, as does the gluten that swells with up to three times its own weight of water. The dough becomes putty-like and un-elastic and comprises some 8% air bubbles. With further mechanical input, the dough is rendered elastic and, if taken to excess, the dough disintegrates.

The rheological changes are primarily determined by the interchange of thiol and disulphide groups in the gluten. The more the cross-liking as disulphide bridges, the firmer the dough. The starch granules become embedded in the matrix and the structure allows gas bubbles to be retained. The pentosans also have a sizeable role in retaining gas by producing a gel-like matrix. This is of particular importance in rye breads when the gluten is of lesser quality.

## Leavening of doughs

Leavened dough absorbs up to three times more heat than does unleavened dough, with the heat penetrating further. In a conventional dough process (with weak gluten flour), the flour, water, salt and yeast are added simultaneously and fermentation is at 26–32°C for only a few hours or perhaps overnight at 18–20°C using less yeast (up to 0.3%). In a sponge dough process (with strong gluten flour), a proportion of the flour, water and yeast are mixed first. After the yeast has multiplied, the remaining materials are mixed in.

There are now continuous processes, and furthermore dough fermentation and maturing may be accelerated by the use of oxidising and reducing agents, the so-called no time doughs. Perhaps the best known of these is the Chorleywood Bread Process developed in 1961. This involves the substitution of biological maturation of dough with mechanical and chemical treatments. The dough is mixed in at high speed (in a 'Tweedy kneader') for 3–5 min under vacuum and in the presence of 75 ppm ascorbic acid. It is also necessary to add fat with a high melting point (approximately 0.7% of the weight of the flour) and more water (ca. 3.5%) to soften the dough for the high mechanical input as well as extra yeast.

## Processing of fermented doughs

Fully fermented dough is divided into pieces that are rounded and allowed to rest for 5–30 min ('intermediate proof'). Their final shape is then established in the moulder, with final leavening in the proof box (30–60 min at 30–40°C) prior to baking.

## Baking

This is of course the most energy intensive stage in the entire process. Temperatures may ordinarily reach 200–250°C for perhaps 50 min for wheat bread. Baking results in a firming or stabilisation of the structure and the formation of characteristic aroma substances. More gas bubbles are generated, leading to an increase in volume of typically 40% and of surface area of 10%.

**Table 12.2**  Flavour components of bread.

| Component | Produced during fermentation | Produced during baking |
|---|---|---|
| Aldehydes | ✓ | ✓ |
| Alkane alcohols | ✓ | |
| Alkene alcohols | | ✓ |
| Amines | ✓ | |
| Diketones | ✓ | |
| Esters | ✓ | |
| Fatty acids | ✓ | |
| Furan derivatives | | ✓ |
| Heterocylic compounds | | ✓ |
| Hydroxy acids | ✓ | |
| Keto acids | ✓ | |
| Ketones | ✓ | ✓ |
| Lactones | ✓ | ✓ |
| Pyrazines | | ✓ |
| Pyridines | | ✓ |
| Pyrroles | | ✓ |

Apart from temperature, the relative humidity in the oven is also important. Firming of the crust must be delayed to permit satisfactory spring and optimal loaf volume. Accordingly, low-pressure steam is directed into the oven at the start of baking and this, by condensing on the surface of the dough, keeps it moist and elastic.

The stages in baking are (1) an enzyme active zone (30–70°C), (2) a starch gelatinisation zone (55°C to <90°C); (3) water evaporation and (4) browning and aroma formation.

## Bread flavour

More than 150 aroma-active substances are generated, including organic acids, their ethyl esters, alcohols, aldehydes, ketones, sulphur-containing compounds, maltol, isomaltol, melanoidin-type substances (in crust) and molecules made by Amadori rearrangements and Strecker degradations. There is also a caramelisation of sugars. Table 12.2 illustrates the contribution that fermentation and baking respectively make to bread flavour.

## Staling of bread

Rather than an overt series of flavour changes (cf. e.g. beer), the staling of bread primarily represents a loss of water. As a consequence, the crumb loses its softness and ability to swell, becoming unelastic, dry and crumbly. Some stale aromas do develop. The physical changes are due to changes in the starch polysaccharides. During baking, amylose diffuses out of granules and, when

**Table 12.3**  Analytical composition of breads.

| | Whole wheat bread | Rye bread |
|---|:---:|:---:|
| Moisture (%) | 40 | 41 |
| Protein (%) | 7.5 | 7.0 |
| Carbohydrate (%) | 49 | 49 |
| Fat (%) | 1.5 | 1.4 |
| Calories (kcal per 100 g) | 240 | 237 |
| Vitamins (% daily need) | | |
| A | 20 | — |
| E | 134 | — |
| $B_1$ | 38 | 28 |
| $B_2$ | 25 | 23 |
| Niacin | 80 | 42 |
| $B_6$ | 60 | — |
| Folic acid | 14 | — |
| Pantothenic acid | 31 | — |
| Minerals (% daily need) | | |
| Calcium | 10–20 | |
| Copper | 50 | |
| Iron | 50 | |
| Magnesium | 70–90 | |
| Manganese | 30 | |
| Phosphorus | 70–80 | |
| Potassium | 60–70 | |

Data from Spicher & Brümmer (1995).

bread cools, this forms a gel that embeds starch granules. Firmness in the crumb is due to heat-reversible association of the side chains of amylopectin within the starch and its retrogradation. Protein and pentosan also seem to be important. Ageing can be minimised by storage at elevated temperatures (45–60°C) or by freezing.

Preservatives such as propionates may be employed to protect against infection with organisms such as *Bacillus mesentericus* (which causes 'rope').

## Bread composition

This is intimately linked to the purity of the flour used to make the bread – that is, the extent to which material has been stripped from the endosperm in milling. Heating, also, will contribute to the loss of materials such as vitamins. The latter may be introduced in fortification treatments, as might minerals and fibre. Data on the analytical composition of breads is given in Table 12.3.

## Bibliography

Cauvain, S.P. & Young, L.S. (1998) *Technology of Breadmaking*. London: Blackie.
Hanneman, L.J. (1980) *Bakery: Bread and Fermented Goods*. London: Heinemann.

Kulp, K. & Ponte, J.G. (2000) *Handbook of Cereal Science and Technology*, 2nd edn. New York: Marcel Dekker.

Spicher, G. & Brümmer, J.-M. (1995) Baked goods. In *Biotechnology*, 2nd edn, vol. 9, Enzymes, Biomass, Food and Feed (eds H.-J. Rehm & G. Reed), pp. 241–319. Weinheim: VCH.

Stauffer, C.E., ed. (1990) *Functional Additives of Bakery Foods*. New York: Van Nostrand Reinhold.

# Chapter 13
# Meat

The curing of meat pre-dates the Romans as an exercise in enhancing meat quality and preserving it.

It comprises lactic fermentation of mixtures of meat, fat, salt, curing agents (either nitrate or nitrite), reducing agents, spices and sugar. Frequently the meat is encased in a tubular form as sausage.

## The role of components of the curing mixture

Salt solubilises the proteins of the muscle as well as increasing the osmotic pressure such that spoilage by bacteria is suppressed. Naturally it enhances flavour. Levels may range from 2% to 3% to as high as 6% to 8%.

The key component is sodium nitrite, which promotes the typical colour of preserved meats through the formation of nitric oxide compounds by reaction with the haem of myoglobin (Fig. 13.1). Furthermore, it contributes to flavour as well as inhibiting the development of pathogens such as *Clostridium botulinum*. The downside is the production of the potentially carcinogenic nitrosamines and so there are legal limits on how much may be used (e.g. 120 ppm for US bacon). Meat typically has a pH of between 5.5 and 6 after rigor mortis is complete. At this pH, nitrite is converted to $N_2O$, which also features in curing. Nitrate may replace nitrite, in which case it is converted to nitrite through the action of bacteria.

Sodium phosphate increases the water-binding capacity of the protein, leading to a stabilisation of the myofibrils. It also binds heavy metals and thus helps protect against the microbes that need those metals.

Sugar is added to counter the salt flavour-wise and is also the carbon and energy source for any microbes necessary for fermentation, for example, those organisms involved in the reduction of nitrate. This sugar will react during any heating stages in Maillard reaction to impact colour and flavour.

Reducing agents, notably ascorbate, reduce nitrite to the nitric oxide that reacts with myoglobin and also helps to suppress the development of nitrosamines.

Binding agents and emulsifiers may be used to improve stability. They may include soy (or hydrolysed soy) starches and carrageenan.

Finally, antioxidants such as BHT and propyl gallate may be added to counter the development of rancidity through lipid oxidation.

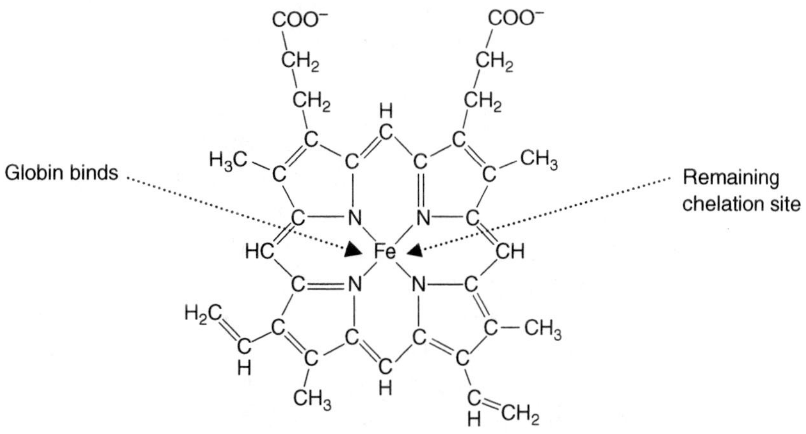

**Fig. 13.1** The interaction of nitrite with haem. The sixth binding site, occupied by nitrite, is the one otherwise occupied by oxygen, carbon monoxide, cyanide, etc.

**Table 13.1** Classifications of fermented sausage.

| Type | $A_w$ | Fermentation time (weeks) | Surface mould growth | Smoked/ not smoked | Example | Origin |
|---|---|---|---|---|---|---|
| Dry | <0.9 | >4 | Yes | No | Salami | Italy |
| Dry | <0.9 | >4 | Yes | Yes | Salami | Hungary |
| Dry | <0.9 | >4 | No | Either | Dauerwurst | Germany |
| Semi-dry | 0.9–0.95 | <4 | Yes | No | Various | France, Spain |
| Semi-dry | 0.9–0.95 | 1.5–3 | No | Usually | Most fermented sausages | Germany, Holland, Scandinavia, USA |
| Undried | 0.9–0.95 | <2 | No | Either | Sobrasada | Spain |

Adapted from Lücke (2003).

# Meat fermentation

The meats are usually classified as either dry or semi-dry (Table 13.1). Dry sausages have an $A_w$ of less than 0.9, tend not to be smoked or heat processed and are generally eaten without cooking. Semi-dry products have an $A_w$ of 0.9–0.95 and are customarily heated at 60–68°C during smoking.

The fermentation temperature is normally below 22°C for dry and mould-ripened sausages, but 22–26°C for semi-dry sausages.

If a starter is used, then the pH reached is in the range of 4–4.5. Starter cultures are primarily the lactic acid bacteria lactobacilli and pediococci, such as *Lactobacillus sakei, Pediococcus pentosaceus, Lactobacillus curvatus, Lactobacillus plantarum* and *Lactobacillus pentosus*. Also of importance, especially when nitrate replaces nitrite, are the non-pathogenic catalase positive cocci *Streptococcus carnosus* and *Micrococcus varians*.

If no starter culture is used, then the pH reaches only 4.6–5. Fermentation here is dependent upon endogenous organisms such as *Lactobacillus sakei* and *Lb. curvatus*.

In the production of fermented sausages, the comminuted lean and fatty tissue is mixed with salt, spice, sugar, curing agent and starter cultures and put into casings. The $A_w$ of a starting semi-dry sausage mix is achieved by employing some 30–35% of fatty tissue and 2.5–3% salt. Nitrite is added in the range of 100–150 mg kg$^{-1}$, and ascorbic acid is also generally included at 300–500 mg kg$^{-1}$. For dry sausages, nitrate may replace nitrite and the fermentation temperature is likely to be lower. Mixes incorporate 0.3% glucose to act as substrate for lactic acid bacteria. The oxygen is rapidly consumed by endogenous meat enzymes. The acid produced in fermentation promotes the reaction of nitrite with metmyoglobin to produce NO-myoglobin. Any residual nitrite is reduced by the microflora. The temperature is lowered to approximately 15°C and the relative humidity in the chamber is brought down to 75–80%. Much of the flavour and aroma that develops is due to the degradation of lipids, notably through autoxidation and the microbial transformation of the products generated by lipid degradation (Fig. 13.2). Additionally, proteinases produce peptides that are converted by the microflora to amino acids and volatile fatty acids.

The sausage may be aged (dried) and smoked. A surface growth may be allowed to develop and this comprises *inter alia* salt-tolerant yeasts (e.g. *Debaromyces hansenii*) and moulds. Where smoking is performed, surface microflora are eliminated. The flora may also be reinforced by starters of *Penicillium nalgiovense* or *Penicillium chrysogenum*. The surface moulds scavenge oxygen and assist the drying process.

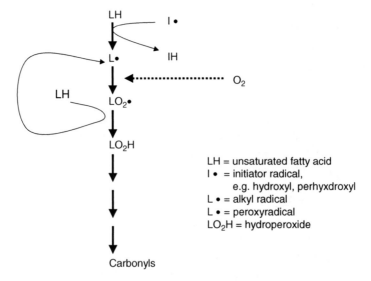

LH = unsaturated fatty acid
I • = initiator radical,
        e.g. hydroxyl, perhyxdroxyl
L • = alkyl radical
L • = peroxyradical
LO$_2$H = hydroperoxide

**Fig. 13.2** The fundamental route for autoxidation of unsaturated fatty acids.

The pH of unground meat must be below 5.8 to prevent the growth of undesirable organisms (pathogens). It is also important that the raw material should not be oxidised (i.e. it should have a low peroxide value). To this end, the meat may first be chilled or frozen to prevent oxidation. Furthermore, the access of oxygen to the meat will be minimised at all stages. To ferment unground meat, salt is first rubbed into the surface, or the meat is dipped in brine, or it is injected with the salt. The meat is then kept at 10°C to allow the salt to become evenly distributed throughout the piece. The meat is then shifted to 15–30°C to allow for water loss and the action of endogenous proteinases in the meat, which degrade the protein structure and increase tenderness and improve the flavour. During this time, a surface bloom of cocci, moulds and yeasts may develop. The meat may be smoked and then dried to the target $A_w$.

## Bibliography

Campbell-Platt, C.H. & Cook, P.E. (1994) *Fermented Meats.* London: Blackie.

Lücke, F.-K. (2003) Fermented meat products. In *Encyclopedia of Food Sciences and Nutrition* (eds B. Caballero, L.C. Trugo & P.M. Finglas), pp. 2338–2344. Oxford: Academic Press.

Varnam, A.H. & Sutherland, J.P. (1995) *Meat and Meat Products.* London: Chapman & Hall.

# Chapter 14
# Indigenous Fermented Foods

A wide diversity of fermented foods that can be pulled together under the generic term 'indigenous' is found worldwide. In Table 14.1, a very few of them are listed and the reader is referred to Campbell-Platt (1987) for a more comprehensive inventory. By way of example, I look in some depth at only three, all from Japan: soy sauce, miso and natto.

## Soy sauce

The history of soy sauce in Japan can be traced back some 3000 years: it probably arrived in Japan from China with the introduction of Buddhism. Although there is an acid-based chemical method for making the product, we focus only at the fermentative route to soy sauce.

Five types of soy sauces are recognised by the Japanese government (Table 14.2). The major types of soy sauce are *Koikuchi*, which accounts for some 90% of the total market and is a multi-purpose seasoning with a strong aroma and a dark red/brown colour, and *Usukauchi*, which is lighter and milder and is employed in cooking when the original food flavour and colour are paramount.

All soy sauces comprise 17–19% salt, seasoning and flavour enhancers.

The overall procedure involved in making soy sauce is given in Fig. 14.1. There are basically two different processes, namely the soaking and cooking of soybeans and the roasting and cracking of wheat.

The soybeans may be whole or the starting material may be de-fatted soybean meal or flakes. When whole beans are used, the oil must ultimately be removed to avoid the production of an unsatisfactory product.

The use of pressed or solvent-extracted meal is less costly and allows a faster, more efficient fermentation due to better access of the relevant enzymes and organisms.

Whole beans or meal are soaked at room temperature (ideally 30°C) for 12–15 h such that there is a doubling of their weight. The water either flows continuously over the beans or is added batch-wise with changes every 2–3 h. This prevents heat accumulation and the development of spore-forming bacteria.

The swollen material is drained, re-covered with water and steamed to induce softening and afford pasteurisation. This is followed by rapid cooling to less than 14°C on 30-cm trays over which air is forced to avoid spoilage.

**Table 14.1**  A selection of indigenous fermented foods (see also Chapter 18).

| Foodstuff | Notes |
| --- | --- |
| Ang kak | Asian colorant based on *Monascus purpureus* growing on rice |
| Bouza | Thick sour wheat-based drink from Egypt |
| Burukutu | Creamy turbid drink in Nigeria made by fermentation of sorghum and cassava by Saccharomyces, Candida and lactic acid bacteria |
| Chichwangue | Bacterial fermentation of cassava root in Congo, eaten as a paste |
| Dosai | Indian spongy breakfast pancake from black gram flour and rice, fermented by yeasts and *Leuconostoc mesenteroides* |
| Idli | Indian bread substitute, also from black gram and rice with fermentation by *Leuconostoc mesenteroides, Torulopsis candida, Trichosporon pullulans* |
| Jalebies | Indian confectionery from wheat flour by *Saccharomyces bayanus* |
| Kaanga-kopuwai | Fermented maize – soft and slimy – eaten as a vegetable |
| Ketjap | Indonesian liquid condiment from fermentation of black soybean by *Aspergillus oryzae* |
| Lao-chao | Glutinous dessert in China from rice fermentation by *Chlamydomucor oryzae, Rhizopus chinensis, Rhizopus oryzae, Saccharomycopsis* sp. |
| Ogi | Breakfast food in Nigeria and West Africa made from corn (maize) – fermentation by lactic acid bacteria, *Aspergillus, Candida, Cephalosporium, Penicillium, Saccharomyces* |
| Poi | Hawaiian side dish to accompany meat and fish made from Taro corms. Relevant organisms: *Candida vini, Geotrichum candidum, Lactobacilli* |
| Rabdi | Semi-solid mush eaten with vegetables in India and made by fermentation of corn and buttermilk |
| Tapé | Soft solid staple fresh dish in Indonesia made from cassava or rice with the aid of *Chlamydomucor oryzae, Emdomycopsis fibuliger, Hansenula anomala, Mucor* sp., *Rhizopus oryzae, Saccharomyces cerevisiae* |

**Table 14.2**  Soy sauces recognised by the Japanese Government.

| Type | Specific gravity (Baumé) | Alcohol (%ABV) | Total nitrogen (g per 100 mL) | Reducing sugar (g per 100 mL) | Colour |
| --- | --- | --- | --- | --- | --- |
| Koikuchi | 22.5 | 2.2 | 1.55 | 3.8 | Deep brown |
| Saishikomi | 26.9 | Trace | 2.39 | 7.5 | Dark brown |
| Shiro | 26.9 | Trace | 0.5 | 20.2 | Yellow/tan |
| Tamari | 29.9 | 0.1 | 2.55 | 5.3 | Dark brown |
| Usukuchi | 22.8 | 0.6 | 1.17 | 5.5 | Light brown |

All the soy sauces have a pH in the range 4.6–4.8 and salt levels between 17.6 and 19.3 g per 100 mL. Derived from Fukushima (1979).

At the same time, wheat (or wheat flour or bran) is roasted to generate the desired flavour characteristics. Products include vanillin and 4-ethylguaiacol from the degradation of lignin and glycosides (Fig. 14.2). The degree of roasting will also impact the colour.

The word 'koji' means 'bloom of mould'. Koji for soy sauce (known as tane) involves the culture of mixed strains of *Aspergillus oryzae* or *Aspergillus sojae* on either steamed polished rice or (less frequently and in China) a mix of wheat bran and soybean flour. It is added to the soybean/wheat mix at 0.1–0.2% to produce koji. Important characteristics of the selected strains

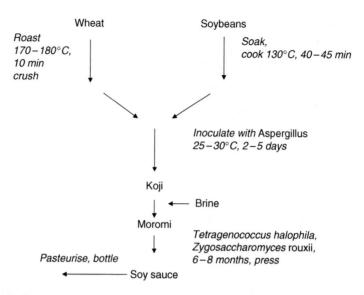

**Fig. 14.1**  Soy sauce production.

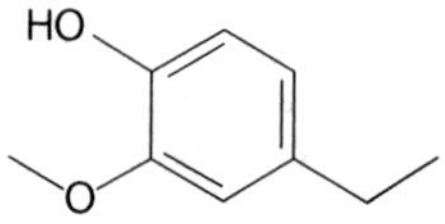

4-Ethylguaiacol

**Fig. 14.2**  A contributor to the flavour of soy sauce.

are the ability to generate high levels of several enzymes (protease, amylase, lipase, cellulase and peptidase) and they should favourably contribute to the aroma and flavour of the final product.

A 1:1 soybean:wheat mixture is spread in 5-cm layers on bamboo (or steel) trays and inoculated with the koji starter. The trays are stacked such that there is good circulation of air, with control of the temperature in the range 25–35°C. Moisture control is important – a high level at first allows mycelial growth, but lower later when the spores are being formed. This stage takes some 2–5 days. Incubation is sufficient for enzymes to be developed, but not too prolonged because otherwise, sporulation occurs, which is accompanied by the development of undesirable flavours.

### Mash (*moromi*) stage

When the koji is mature, it is mixed with an equal volume of saline, with the target sodium chloride level being 17–19%. Less than that allows the development of putrefactive organisms. If the salt content is too high, there is an inhibition of desirable osmophilic and halophilic organisms. The salt destroys the koji mycelium.

Originally (and still at the craft level) fermentation is not temperature regulated and can take 12–14 months. On a commercial scale in wood or concrete fermenters, the temperature is controlled to 35–40°C for a period of 2–4 months. The must is mixed from time to time with a wooden stick on the small scale or with compressed air on the large scale. The enzymes of the koji hydrolyse proteins early in the fermentation process to generate peptides and amino acids. Then the amylases release sugars from starch, these being fermented to lactic, glutamic and other acids, causing the pH to fall to 4.5–4.8. Carbon dioxide is also produced. If this is in excess, then there is too much opportunity for anaerobic organisms to develop, with attendant flavour difficulties. Conversely, if there is excessive oxygen, then the fermentation does not proceed according to the desirable course.

The microbiology of soy sauce production is not fully appreciated. In the earliest stages, halophilic *Pediococcus halophilus* predominates, converting sugars to lactic acid and dropping pH; followed by *Zygosaccharomyces rouxii, Torulopsis* and certain other yeasts.

Table 14.3 lists some of the compounds that contribute to the flavour of soy sauce. Yeasts make the biggest contribution to the flavour of soy sauce, generating *inter alia* 4-ethyl guiaicol, 4-ethylphenol, ethanol, pyrazones, furanones, ethyl acetate. Acids are generated by Pediococcus and perhaps lactic acid bacteria.

**Table 14.3**   Some of the compounds that contribute to the flavour of soy sauce.

| | |
|---|---|
| Acetaldehyde | Furfural |
| Acetic acid | Furfuryl acetate |
| Acetoin | Furfuryl alcohol |
| Acetone | Guaiacol |
| 2-Acetyl furan | 2,3-Hexanedione |
| 2-Acetyl pyrrole | 2-Hexanone |
| Benzaldehyde | 4-Hydroxy-2-ethyl-5-methyl-3(2H)-furanone |
| Benzoic acid | 4-Hydroxy-5-ethyl-2-methyl-3(2H)-furanone |
| Benzyl alcohol | 4-Hydroxy-5-methyl-3(2H)-furanone |
| Borneol | Maltol |
| Bornyl acetate | Methional |
| Butanoic acid | 3-Methylbutanal |
| 1-Butanol | 3-Methylbutanoic acid |
| Diethyl succinate | 3-Methyl-1-butanol |
| 2,6-Dimethoxyphenol | 3-Methylbutyl acetate |
| 2,3-Dimethylpyrazine | 2-Methylpropanal |
| 2,6-Dimethylpyrazine | 2-Methylpropanoic acid |
| Ethanol | 2-Methyl-1-propanol |
| Ethyl acetate | 3-Methylpyrazine 3-methyl-3-tetrahydrofuranone |
| Ethyl benzoate | 4-Pentanolide |
| 3-Ethyl-2,5-dimethylpyrazin | Phenylacetaldehydee |
| 4-Ethylguaiacol | 2-Phenylethanol |
| Ethyl lactate | 2-Phenylethylacetate |
| 2-Ethyl-6-methylpyrazine | Propanal |
| Ethyl myristate | 2-Propanol |
| 4-Ethylphenol | |
| Ethyl phenylacetate | |

Liquid is removed from the mash by pressing (hydraulic presses are used in large-scale operations) and new salt water may be added to the residue. A second fermentation may proceed for 1–2 months generating a lower quality product. Oil is removed from filtrate by decantation.

Raw soy sauce is pasteurised at 70–80°C to kill vegetative cells and denature enzymes. Alum or kaolin may be added as clarifiers before the product is filtered and bottled. Para-hydroxybenzoate or sodium benzoate may be added as antimicrobials.

## Miso

There are various fermented soybean pastes in Asia, including miso in Japan, Chiang in China, Jiang in Korea, Tauco in Indonesia, Taochieo in Thailand and Taosi in the Philippines.

Miso, nowadays made commercially, is for the most part used as the base for soups, with the remainder being employed in the seasoning of other foods.

There are four basic steps, two of which are concurrent, namely the preparation of koji and of soybeans.

Koji is made on polished rice and represents a source of enzymes that will hydrolyse soybean components. Waxy components in the outer layers of unpolished rice inhibit the penetration by the Aspergillus mycelium. The rice is washed and soaked overnight at 15°C to a moisture content of 35%. Excess water is removed and the material is steamed for 40–60 min. The rice is then spread on large trays and cooled to 35°C. Seed koji (see the section on soy sauce) is added at 1 g per kg rice.

The trays in koji rooms tend nowadays to be replaced by rotary drum fermenters that facilitate control of temperature, air circulation and relative humidity, as well as avoiding agglomeration of the rice. The temperature is held to 30–35°C over a period of 40–50 h. In this time, the rice becomes covered with white mycelium. Harvesting occurs before the occurrence of sporulation and pigment development. The material has a sweet aroma and flavour.

Salt is added as the material is removed from the fermenter so as to prevent further microbial growth.

The whole soybeans employed for miso are large and selected for their ability to absorb water and cook rapidly. They are washed before soaking for 18–22 h. The water is changed regularly especially during summer months in order to prevent bacterial spoilage. The beans swell to almost 2.5 times their volume. After draining, the beans are steamed at 115°C for 20 min when they become compressible.

The beans are mixed with salted koji. Starter cultures may be introduced, including osmophilic yeasts and bacteria. The microflora includes *Z. rouxii*, Torulopsis, Pediococcus, Halophilus and *Streptococcus faecalis*.

The mixture, known as 'green miso', is packed into vats and anaerobic fermentation and ageing are allowed to proceed at 25–30°C for various periods depending on the character required. Transfer occurs between vessels at least twice. White miso takes 1 week, salty miso 1–3 months and soybean miso over 1 year. The miso is blended, mashed, pasteurised and packaged.

The characteristics of different miso are listed in Table 14.4.

Amino acids represent a significant source of miso flavour, and they are generated from soybean protein by the action of proteinases (which may be supplemented from exogenous sources). Miso contains 0.6–1.5% acids (lactic, succinic and acetic) as a result of sugar fermentation. Esters produced from the reaction of alcohols with some fatty acids from the soybean lipid are also important flavour contributors.

## Natto

Natto is a Japanese product based on fermented whole soybeans. Generally the product is dark with a pungent and harsh character. It is eaten with boiled rice, as a seasoning or as a table condiment in the way of mustard.

There are three types of natto in Japan.

*Itohiki-natto* from Eastern Japan is produced by soaking washed soybean overnight to double its weight, steaming for 15 min and inoculating with *Bacillus natto*, which is a variant of *Bacillus subtilis*. Fermentation is allowed to proceed for 18–20 h at 40–45°C. Polymers of glutamic acid are produced which afford a viscous surface and texture in the final product.

*Yuki-wari-natto* is produced by mixing itohiki-natto with salt and rice koji and leaving at 25–35°C for 2 weeks.

For *hama-natto*, soybeans are soaked in water for 4 h and steamed for 1 h, before inoculating with koji from roasted wheat and barley. After 20 h (or when covered with green mycelium of *A. oryzae*), the material is either sun-dried or dried by warm air to about 12% moisture. The beans are submerged in salt brine containing strips of ginger and allowed to age under pressure for

**Table 14.4**  Types of miso.

| Base material | Colour | Taste | Time of fermentation/ ageing |
| --- | --- | --- | --- |
| Rice | Yellow-white | Sweet | 5–20 days |
| Rice | Red-brown | Sweet | 5–20 days |
| Rice | Light yellow | Semi-sweet | 5–20 days |
| Rice | Red-brown | Semi-sweet | 3–6 months |
| Rice | Light yellow | Salty | 2–6 months |
| Rice | Red-brown | Salty | 3–12 months |
| Soybeans | Dark red-brown | Salty | 5–20 months |
| Barley | Yellow-red-brown | Semi-sweet | 1–3 months |
| Barley | Red-brown | Salty | 3–12 months |

Based on Fukushima (1979).

up to 1 year. The surface microflora contributing to enzymolysis and flavour development includes Pediococci, Streptococci and Micrococci.

# Bibliography

Beuchat, L.R. (1987) *Food and Beverage Mycology*, 2nd edn. New York: Van Nostrand Reinhold.

Campbell-Platt, G. (1987) *Fermented Foods of the World: A Dictionary and Guide.* London: Butterworths.

Fukushima, D. (1979) Fermented vegetable (soybean) protein and related foods of Japan and China. *Journal of the American Oil Chemists' Society*, **56**, 357–362.

Reddy, N.R., Pierson, M.D. & Salunkhe, D.K., eds (1986) *Legume-based Fermented Foods.* Boca Raton: CRC Press.

Steinkraus, K.H. (1996) *Handbook of Indigenous Fermented Foods*, 2nd edn. New York: Marcel Dekker.

# Chapter 15
# Vegetable Fermentations

The pickling of vegetables for purposes of preservation probably originated in China with the use of brines, and subsequently dry salting. The three vegetables of most commercial significance in this context are cabbage, cucumbers and olives, but others that may be fermented include artichokes, beet, carrots, cauliflower, celery, garlic, green beans, green tomatoes, peppers, turnip and a variety of Asian commodities (see Kimchi in Chapter 18).

## Cucumbers

Whereas cucumbers (*Cucumis sativus*) retailed for their direct use are customarily bred to have tough skins, those targeted for pickling need to have a thin and relatively tender coating. They are harvested at a relatively immature stage, before the seeds have matured and before the area around the seeds has gone soft and starts to liquefy through the action of polygalacturonases on cell-wall hemicelluloses. The most valuable cucumbers are also the smaller ones. The cucumbers are sorted according to their diameter, and those that are too long are cut to a length that will readily fit into jars.

Other breeding criteria include disease resistance, yield, the growth locale and a relatively small seed area. Cucumbers should be straight and uniform with a length to diameter ratio of 3 : 1. They should be firm, green and free from internal defects. Chemical parameters include the level of cucurbitacins, which afford bitterness, sugars (which are the substrates for the fermentation), malic acid (relevant to the extent to which 'bloaters' are produced during fermentation) and the level of polygalacturonase. Opportunities for molecular biology in the optimisation of these parameters are being explored.

Cucumbers that are grown locally are processed within 1 day, whereas those grown further afield are refrigerated on shipping. If brined, they can be transported internationally.

Pickling cucumbers are preserved by one of three methods. Some two-fifths are preserved by fermentation, possibly accompanied by pasteurisation. Pasteurisation alone (reaching an internal temperature of 74°C for 15 min) is applied to another 40% of the total, while the remainder rely solely on refrigeration. For pasteurised and refrigerated processing, acid (produced separately, i.e. not through *in situ* fermentation) is usually added, perhaps accompanied by sodium benzoate.

**Table 15.1**  Stages of microbial involvement in vegetable fermentation.

| Stage | Microbial events |
|---|---|
| Start | A range of Gram-positive and Gram-negative bacteria present |
| Primary fermentation | Most bacteria inhibited in the acid conditions created by the lactic acid bacteria. Lactic acid bacteria and yeast are able to thrive |
| Secondary fermentation | Lower pH now inhibiting lactic acid bacteria, but not yeasts growing fermentatively |
| Post-fermentation | Surface growth of oxidative bacteria, moulds and yeasts in open tanks. However, if in sealed anaerobic tanks, no growth if pH is low enough and salt concentration high enough |

Based on Fleming (1982).

Most commercial cucumber fermentations rely on a natural microflora. Sometimes, however, the natural microflora is heavily depleted by hot water blanching (66–80°C for 5 min), in which case there may be seeding with *Lactobacillus plantarum*. The various stages of microbial growth are indicated in Table 15.1. When the flower has withered, it tends to have increased levels of micro-organisms and, furthermore, the flowers also contain polygalacturonase that plays a significant role in softening cucumbers by hydrolysing the polysaccharide matrix. The major fermentation sugars are glucose and fructose and these are metabolised to lactic acid, acetic acid, ethanol, mannitol and carbon dioxide. *Lb. plantarum* is normally the predominant organism in the natural microflora, mostly producing lactic acid. A malolactic fermentation is important in converting malate in the cucumbers to lactic acid.

The fresh cucumbers are immersed in brine in bulk tanks. The control parameters are pH, temperature and the level of salt. The brine is typically lowered to a pH of around 4.5 with either vinegar or lactic acid. This facilitates the loss of carbon dioxide (by shifting the equilibrium from bicarbonate towards carbonic acid). Furthermore, it has a major impact on which organisms grow, for instance, the growth of Enterobacteriaceae is suppressed at the lower pH whereas lactic acid bacteria are able to thrive in the absence of competition from organisms not able to tolerate these acidic conditions. The optimum salt level is 5–8% sodium chloride with the temperature in the range 15–32°C. The species involved are listed in Table 15.2.

During fermentation, the brine is purged with either nitrogen or air to prevent bloater formation, and the cucumbers are maintained submerged. Whereas air is the cheaper option, nitrogen is preferable as there is then less yeast and fungal growth, fewer off flavours and less colour development. Potassium sorbate (0.035%) is typically added to inhibit the growth of fungi. It is critical that the end product should possess a firm, crisp texture. Furthermore, as lactic acid is deemed too tart for products such as hamburger dill, a draining stage is employed with replacement of the brine by vinegar.

Pasteurised products typically contain 0.5–0.6% acetic at a pH of 3.7. The relative content of acid and sugar is adjusted depending on the desired sourness/sweetness balance.

**Table 15.2**   Lactic acid bacteria involved in fermentation of vegetables.

Homofermentative
  *Enterococcus faecalis*
  *Lactobacillus bavaricus*
  *Lactococcus lactis*
  *Pediococcus pentosaceus*
Heterofermentative
  *Lactobacillus brevis*
  *Leuconostoc mesenteroides*
Mix
  *Lactobacillus plantarum*[a]

[a] This organism uses hexoses homofermentatively but pentoses heterofermentatively.

## Cabbage

Sauerkraut is pickled cabbage (*Brassica oleracea*). The cabbages of choice will have large heads (8–12 lb) that are compact (dense), contain few outer green leaves and have desirable flavour, colour and texture. They are bred for yield, pest resistance, storability and content of dry matter.

Cabbages are increasingly harvested mechanically and are graded, cored, trimmed, shredded and salted. Their water content is about 30% and shredding is to a diameter of approximately 1 mm.

The shredded cabbage is soaked in brine in reinforced concrete tanks of capacity 20–180 tons and loosely covered with plastic sheeting. Alternatively, cabbage may be dry salted to about 2% by weight and allowed to self-brine through its own moisture. The cabbage is distributed to a slight concave surface and water put on top of the plastic cover to anchor it and ensure that anaerobic conditions can develop. Fermentation can take some 3 weeks, ideally at temperatures below 20°C.

Lactic acid bacteria constitute a relatively small proportion of the total bacterial count and comprise five major species: *Enterococcus faecalis, Leuconostoc mesenteroides, Lactobacillus brevis, Pediococcus cerevisiae* and *Lb. plantarum*. Despite their low levels, these organisms represent the most significant contributor to the fermentation. A low salt concentration (ca. 2%) and the low temperature (18°C) favour heterofermentative organisms. Conversely, a high salt content (3.5%) and high temperature (32°C) promote homofermentative fermentation. The normal sequence is heterofermentation first, followed by homofermentation. The main sugars in cabbage are glucose and fructose and, to a lesser extent, sucrose. They are converted to acetic acid, mannitol and ethanol in the first week, together with $CO_2$ which is important for establishing anaerobiosis. After a week or so, the brine becomes too acidic for the heterofermentative organisms and the fermentation is continued by the homofermenters, notably *Lb. plantarum*. Production of lactic acid continues until all the sugars are consumed and the pH has dropped from around 6 to 3.4.

The cabbage stays in the tanks until more than 1% lactic acid has been produced (30 days or more). The material is then either stored in the same vessel or is processed at this stage to the finished product.

The sauerkraut is removed either manually or by mechanical fork and is packaged into can, glass or plastic. Sodium benzoate (0.1% w/v) may be added as a preservative and the material stored at 4°C. If canned, the product is pasteurised and no preservative is added. Pasteurisation is at 74–82°C for 3 min. Heating is by steam injection or immersion and the product hot filled into cans.

Sauerkraut can be spoiled by Clostridia if the latter proliferates in the early stages of the process. Other potential problem organisms are oxidative yeasts and moulds. Discoloration may arise not only from the oxidation of cabbage components but also from the action of Rhodotorula which generates a red hue.

# Olives

Olives (*Olea europaea*) are primarily fermented in the Mediterranean countries of Greece, Italy, Morocco, Spain and Italy. Part of the reason for the process is to eliminate the acute bitterness of the olive that is due to the glycoside oleuropin. Soaking the olive in brine or dilute caustic leads to the hydrolysis and removal of this material.

Nowadays olives are mostly fermented in plastic-clad tanks of fibreglass or stainless steel, perhaps buried underground in the interests of temperature regulation. There are basically two fermentation approaches.

## Untreated naturally ripe black olives in brine

The olives are picked when completely ripened (turned from green to black or purple) and are not treated with lye (alkali solution) so that they retain bitterness and fruitiness. They are put into the tanks with 6–10% sodium chloride solution and allowed to undergo spontaneous fermentation by an endogenous microflora comprising lactic acid bacteria and yeasts. The olives are subsequently sorted and graded before packaging.

## Lye-treated green olives in brine

The olives are harvested when green or yellow and treated with a 1.3–3.5% lye solution for up to 12 h at 12–20°C to remove most of the bitterness. After washing with cold water, they are taken in stages up to a concentration of 10–13% sodium chloride, a gradual process so as to avoid shrivelling. Endogenous fermentation is allowed to progress for up to a month at 24–27°C, prior to sorting and grading and packaging into glass jars.

In olive fermentations there is no use of starter cultures, although a proportion of brine from a previous fermentation may be used to supplement the new brine.

In the early stages of fermentation, there is activity of the aerobic organisms Citrobacter, Enterobacter, Escherichia, Flavobacterium, Klebsiella and Pseudomonas. These organisms will not grow when the salt is increased beyond 6–10%. Stage two comprises the activity of the lactic acid bacteria (Lactobacillus, Lactococcus, Leuconostoc, Pediococcus), with the progressively dropping pH destroying the initial microflora. The onset of the third stage is once the pH reaches 4.5, with the predominant organism being *Lb. plantarum*, together with fermentative and oxidative yeasts (Candida, Hansenula, Saccharomyces).

## Bibliography

Eskin, N.A.M., ed. (1989) *Quality and Preservation of Vegetables*. Boca Raton: CRC.
Fleming, H.P. (1982) Fermented vegetables. In *Economic Microbiology* (ed. A.H. Rose), pp. 227–258. London: Academic Press.
McNair, J.K., ed. (1975) *All About Pickling*. San Francisco: Ortho.

# Chapter 16
# Cocoa

The starting material for cocoa and chocolate is the seed of *Theobroma cacao* which was first cultivated by the Aztec and Mayan civilisations more than 2500 years ago and imported by the Spanish in 1528. Processing is in the tropics where the cocoa is grown, with ensuing manufacturing in the countries where the end products are consumed.

There are two major types of *T. cacao*. Criollo affords cocoas that have a refined flavour but low yield. Forastero affords much higher yields and is therefore the predominant type used, accounting for approximately 95% of the cocoa beans used in the manufacture of chocolate and cocoa products.

Cocoa pods (Fig. 16.1) develop on the trunks and branches of the tree and are harvested throughout the year. They comprise an embryo and shell. There are between 35 and 45 seeds (or beans or cotyledons) encased in a mucilaginous pulp known as the endocarp and composed of sugars (mainly sucrose), pectins, polysaccharides, proteins, organic acids and salts (Table 16.1). The plant contains alkaloids, notably the methylxanthines theobromine (1–2% of the dry weight) and caffeine (0–2%) (Fig. 16.2) The former affords bitterness to cocoa. The embryo of the seed comprises two folded cotyledons that are covered with a rudimentary endosperm. It is these cotyledons that are used for making cocoa and chocolate (Fig. 16.3).

The ripe pods are harvested and their husks broken using sharp objects or wooden billets. The wet beans are removed from the husk and heaped (50–80 cm deep) on the ground or in boxes (100 cm deep) to allow 'sweatings' to drain from the bottom. The beans are covered mainly with banana leaves, and left for 5–7 days with one or more turnings to allow for a more even fermentation. The temperature will rise to around 50°C and must be maintained below 60°C to avoid over-fermentation and excessive growth of fungi.

During fermentation, the pulp becomes infected with diverse micro-organisms from the environment. At the start of fermentation, the low pH and high sugar in the surrounding pulp favour anaerobic fermentation by yeasts and also the growth of lactic acid bacteria. The ethanol produced represents a substrate for the acetic acid bacteria, which predominate when the sugars are exhausted. Pectinolytic activity is supplied by *Kluyveromyces marxianus,* but Saccharomyces, Torulopsis and Candida are other yeasts that have significant roles to play. The pectinolysis leads to the draining of the pulp off the beans as 'sweatings'. This allows air into the spaces between the beans and so, late in fermentation, aerophiles develop, including Bacillus, as well as filamentous fungi, such as *Aspergillus fumigatus,* Penicillium and Mucor spp.

**Fig. 16.1**   The cocoa pod. Photograph supplied by Dave Zuber of Mars, Incorporated.

**Table 16.1**   The composition of the cocoa cotyledon.

| Component | Percentage by weight |
|---|---|
| Water | 32–39 |
| Cocoa butter (lipid) | 30–32 |
| Protein | 8–10 |
| Polyphenols | 5–6 |
| Starch | 4–6 |
| Pentosans | 4–6 |
| Cellulose | 2–3 |
| Theobromine | 2–3 |
| Salts | 2–3 |
| Sucrose | 2–3 |
| Caffeine | 1 |
| Acids | 1 |

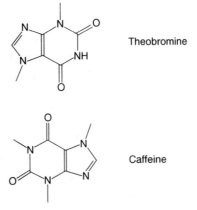

Theobromine

Caffeine

**Fig. 16.2**   Methylxanthines in cocoa.

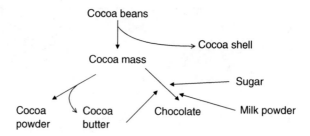

**Fig. 16.3**   An overview of cocoa processing.

The increasing concentration of ethanol and acetic acid, together with a rise in heat, eventually leads to the death of the bean. Once this occurs, the biological barriers within the cotyledon are broken down, permitting the release of several types of enzymes.

Initially the anaerobic conditions inside the cotyledon favour hydrolytic enzymatic reactions but, later, aerobic conditions prevail, which favour oxidative reactions, especially of the polyphenols.

Invertase hydrolyses sucrose to the reducing sugars, glucose and fructose. These will later combine with peptides and amino acids. During roasting of the beans (discussed later), these compounds enter into the Maillard reaction, and the resultant flavoursome substances are highly significant for the flavour of chocolate.

Glycosidases release polyphenols from their attachment to sugars. The anthocyanidins released polymerise to leucocyanidins, which in turn complex with some of the protein, lessening their astringency and bitterness, as well as reducing the levels of unpleasant flavours and odours sometimes associated with roasted proteins.

After fermentation, the beans are exposed to drying, either by sun or by a forced hot-air source. Drying is an important continuation of the fermentation process and, consequently, flavour-precursor development. During drying, aerobic conditions prevail, favouring oxidative reactions, especially of the polyphenols through the action of PPOs. Since fermentation is a gradual process spread over a 5–7-day period, the action of PPO commences towards the end of the anaerobic phase of fermentation. Quinones are also formed by the oxidative changes brought by the action of the PPO on the polyphenols. These complex with free amino and imino groups of proteins, the tanning of the protein leading to a colour change in the beans and a reduction of astringency.

There appears to be a fine balance between the fermentation and drying that must be adhered to if a consistent flavour is to be achieved in the bean. It is barely credible that the crude and sometimes haphazard methods employed allow this balance to be maintained. Care must be taken not to dry the beans too rapidly, which can lead to case hardening of the bean, thus entrapping more of the unwanted volatile acetic acid.

Whichever method is used, it is essential that the beans are dried down to 5–7% moisture to inhibit the development of mould during storage. The ensuing mouldy taste in the chocolate is almost impossible to eradicate by further processing.

The extent to which the biochemical changes have progressed during fermentation and drying is assessed from the colour change of the cotyledons, resulting from the oxidation of the polyphenolic constituents. A brown colour in the bean is indicative of complete fermentation, purple/brown suggests partial fermentation, purple signifies under-fermented and slate-colouration indicates that the bean has not been fermented. Chocolate made from slate-coloured beans is bitter, astringent and almost devoid of chocolate flavour.

Acetic acid is a by-product of the fermentation of the sugars occurring in the surrounding pulp and significant diffusion into the cotyledon during fermentation causes a decrease in the pH of the beans. For some types of Forastero beans, pH is used as a secondary measurement of the degree of fermentation.

Levels of theobromine and caffeine decline during fermentation, as is also the case for the lipid component of the bean, cocoa butter.

Cocoa butter is fully saturated, hence it is one of the most stable fats in nature and resistant to oxidation. Depending on its polymorph, cocoa butter has a melting temperature of approximately 34.5°C, some 2.5°C lower than normal body temperature. Its melting profile is sharp, so that the chocolate made from it melts cleanly in the mouth with no residual, waxy aftertaste. However, sufficient unmelted solids remain to give body to the chocolate at regular distribution temperatures.

The melt temperature of cocoa butter varies according to the genetics and geographical source of the cocoa. Malaysian cocoa butter has the highest melt temperature and is the hardest in texture. Depending on the season, Brazilian cocoa butter, produced from the winter crop, is the softest and has the lowest melting temperature.

The starch remains virtually chemically unchanged during the fermentation process.

# Roasting

Roasting results in the reduction of moisture in the beans from 7% to approximately 1.5%. Much of the volatile acidity, mainly acetic, is evaporated. Non-enzymatic browning and Strecker reactions occur, leading to a diversity of molecules that represent the main part of the chocolate flavour and aroma. These include several types of pyrazines, aldehydes, ketones, esters and oxazoles. Some 400–500 compounds form the basis of chocolate flavour.

Depending on the geographical origin of the beans, roasting temperatures will vary between 110°C and 220°C. The lower temperatures are used for the more fragile and subtly flavoured beans.

## Production of cocoa mass or chocolate liquor

At the beginning of the process of converting the dried cocoa beans into chocolate liquor, the beans are first passed over magnets and vibrating screens to remove any unwanted debris. The beans are then roasted whole, then winnowed or passed over infrared heaters to pop the outer shell. This shell is then removed by a winnowing process which separates the non-usable shell from the nib (raw cotyledon).

The roasted, de-shelled bean and/or nibs are ground to a fine particle size of about 100–120 μm by different types of grinding machines, such as stone, ball, pin mills, etc.

## Cocoa butter

This is extracted from the milled chocolate liquor by mechanical pressing through mesh metal screens by hydraulic presses operating at high pressure at about 90°C. The resulting cocoa butter has a distinct chocolate flavour, which some companies deem too strong for milk chocolate. They prefer to use a more odourless steam-deodorised cocoa butter.

A by-product of pressing the chocolate liquor is cocoa press cake and this is pulverised to cocoa powder.

Depending on the pressure that the chocolate liquor has been exposed to, the residual cocoa butter content of the cocoa powder ranges from 10% to 20%. Defatted cocoas are processed either by expeller press or solvent extraction.

## Production of chocolate

Sugar (usually pre-pulverised), chocolate liquor and whole milk powder are first mixed to form a paste that can be passed through a five-roll refiner. The paste is ground to an average particle size, which for regular commercial chocolate is about 10–15 μm.

This paste is filled into a machine known as a conche, within which there is dry mixing and aeration on a massive scale. During the conching process, which can take between 6 and 72 h, the moisture and volatile acids are evaporated which results in a reduction of the viscosity of the chocolate. For milk chocolate, conching is performed at 50–65°C, but for dark chocolate it is in the range 60–90°C.

Due to the high shearing forces for long periods in the conching process, major changes occur in the texture. The finished chocolate is more cohesive, less crumbly when set, and the taste is much more mellow and less harsh and bitter. The loss of acetic acid ensures a reduction in acid taste. Chocolate receiving high-shearing action and, therefore, better aeration, shows a reduction in astringency, which would suggest that further oxidation of polyphenols is occurring.

During lengthy shearing, there is a better distribution of fat over the dry particles, especially the highly flavoured 'spikey' particles. This may result in a smoother, less bitter astringent taste in the finished chocolate.

The final step in the conching process is the addition of lecithin to reduce the viscosity of chocolate to a workable rheological mass.

The chocolate is now ready for use in either a coating or moulding operation.

Cocoa butter has five distinct polymorphs and, before it can be used in coating or moulding, it must be put through a cooling, mixing regime to achieve the correct stable form V polymorph. This process is called tempering. There are literally dozens of ways to achieve the correct stable cocoa butter crystallisation.

Tempering involves first cooling the chocolate with agitation, taking the temperature from 45–50°C to approximately 27–28°C. At this point, the chocolate is quite viscous and will contain the unstable form IV polymorph. The temperature is then raised to a working temperature of between 29°C and 32.5°C, which will vary depending on the source of cocoa butter and the presence of anhydrous dairy butter fat. After coating and moulding, the chocolate must be carefully cooled to avoid the re-introduction of form IV crystals. The chocolate is now ready for packing and is preferably held at a constant 18°C during the distribution cycle.

## Bibliography

Beckett, S.T. (1988) *Industrial Chocolate Manufacture and Use.* London: Blackie.

Cook, L.R. & Meursing, E.H. (1982) *Chocolate Production and Use.* New York: Harcourt Brace Jovanovich.

Dimick, P.S., ed. (1986) *Proceedings of Cocoa Biotechnology.* Philadelphia: Pennsylvania State University.

Richardson, T.W. (2000) Back to basics – chocolate tempering. *Proceedings of the PMCA Production Conference* (http://pmca.com/).

Wood, G.A.R. & Lass, R.A. (1985) *Cocoa,* 4th edn. Harlow: Longman.

# Chapter 17
# Mycoprotein

Although less high profile than it was 25–30 years ago, there is still interest in the cultivation of microbes specifically as foodstuffs, rather than as agents in the production of other products, which is how we have encountered them in this book. The term 'single cell protein' was coined to describe these products, which were based on diverse bacteria and yeasts, growing on a range of carbon sources (Table 17.1).

Only one product has survived in substantial quantity to this day, Quorn™. It is a joint venture between two major British companies and has been marketed as a meat substitute since 1984.

The organism, *Fusarium venenatum*, is grown at 30°C in rigorously sterile conditions in air lift (pressure cycle) fermenters. The liquid medium flows continuously into the fermenter (the residence time is 5–6 h), and the conditions are highly aerobic, with the compressed air serving both as nutrient and as the vehicle for agitation.

Carbon source is glucose produced by the hydrolysis of corn starch, and ammonium salts are included as the nitrogen source. The pH is maintained at 4.5–7.0 and iron, manganese, potassium, calcium, magnesium, cobalt, copper and biotin are added. Unlike the other products considered in this book, the cells themselves are really all that impact on the properties of the finished product in the present instance. The medium composition is relevant only

Table 17.1  Some single cell protein processes.

| Substrate | Organism |
|---|---|
| Cellulose | *Alcaligenes, Cellulomonas* |
| Ethanol | *Candida utilis, Acinetobacter calcoaceticus* |
| Glucose | *Fusarium venenatum* |
| Hydrocarbons | *Candida tropicalis, Yarrowia lipolytica* |
| Methane | *Methylococcus capsulatus* |
| Methanol | *Methylomonas clara, Methylophilus methylotrophus, Pichia pastoris* |
| Molasses | *Candida utilis* |
| Starch | *Saccharomyces cerevisiae, Saccharomycopsis fibuligera/Candida utilis* |
| Sucrose | *Candida utilis* |
| Sulphite waste liquor | *Candida utilis* |
| Whey | *Candida intermedia, Candida krusei, Candida pintolepesii, Candida utilis, Kluyveromyces lactis, Kluyveromyces marxianus, Lactobacillus bulgaricus* |

insofar as it impacts the yield and properties of the organism *per se* and has no role to play, for instance, in determining final product flavour or appearance.

The continuous fermentation system will be re-established every 1000 h.

After fermentation, the cell suspension is heat-shocked to reduce the extent of development of RNA degradation products, the presence of which will otherwise elevate the risk of gout in those partaking of the foodstuff. Heating is at 64°C to eliminate the enzymes that convert RNA to nucleotides.

The cell suspension is harvested by centrifugation and the hyphae mixed with binding agents and flavourants and heated to cause a gelling of the binder and a linking of the hyphae.

The product is some 45% protein, 14% fat and 26% fibre by dry weight. It is 11% protein, 3% available carbohydrate, 6% fibre, 3% fat, 2% ash and 75% water by wet weight. It is sold in a variety of commercial forms, for example, pieces and minced.

Nutritionally, it stacks up very well against other foods. It possesses a complete complement of essential amino acids and is a particularly good source of threonine, which tends to be the limiting amino acid in meat. Quorn has little saturated fat and has a favourable ratio of polyunsaturated to saturated fatty acids when compared with beef and chicken. It is devoid of cholesterol and is low in calories. It possesses significant levels of fibre in the form of chitin and $\beta$-glucan from the Fusarium cell walls. It contains the breadth of B vitamins, with the exception of $B_{12}$. Finally it is devoid of phytic acid, and so tends not to interfere with metal uptake from the diet.

## Bibliography

Goldberg, I. (1985) *Single Cell Protein*. Berlin: Springer-Verlag.

Large, P.J. & Bamforth, C.W. (1988) *Methylotrophy and Biotechnology*. London: Longman.

Moo-Young, M. & Gregory, K. (1986) *Microbial Biomass Proteins*. London: Elsevier.

Tanenbaum, S.R. & Wang, D.I.C., eds (1974) *Single Cell Protein II*. Cambridge, MA: MIT Press.

Trinci, P.J. (1991) Quorn mycoprotein. *Mycologist*, **5**, 106–109.

Wainwright, M. (1992) *An Introduction to Fungal Biotechnology*. Chichester: Wiley.

# Chapter 18
# Miscellaneous Fermentation Products

**Table 18.1**

| Foodstuff | Details | Origin |
|---|---|---|
| Acidophilus milk | Skim or full fat milk, sterilised, incubated with *Lactobacillus acidophilus* or *Bifidobacterium bifidum* (<48 h). Therapeutic value: lowering pH of intestine | Europe and North America |
| Apéritif wine | Bitter tasting, high alcohol wine, often red, drunk before meals. Red wine or white wine strengthened with added grape spirit or alcohol, flavourings. For example, Campari from Italy = red and flavoured with quinine. Dubonnet – France = red or white, flavoured with quinine and herbs | International |
| Bacon (see also Chapter 13) | Pork sides cured – curing salts containing some or all of sodium chloride, potassium nitrate, sodium nitrite, sugars, ascorbic acid. Covered in curing pickle – 3–6°C for 2–10 days. Taken away from brine and stored at same temperature for up to 2 weeks. May be cold smoked at 25–35°C or cooked to internal temperature of 50–55°C. Bacteria – Micrococcus or Staphylococcus – reduce nitrate to nitrite, which is active form in producing active pink nitroso compounds. Lactobacillus active in maturing. Shorter process may find chemical curing more important than microbial curing | International |
| Bagel (see also Chapter 12) | Traditional Jewish bread. Baker's yeast and sometimes egg added to wheat flour dough, fermenting and proofing 40–50 min, knocked back to original size by expelling gas, dividing and rolling into balls, grilled 4–5 min at 200°C, dropped into boiling water for 15–20 min, drained and baked in oven at 200°C for 15 min until crisp | Middle East, North America |
| Bagoong | Fermented salty fish paste. Condiment with rice dishes in Asia. Remove heads and eviscerate fish. May be sun dried for 3–4 days and then pounded. One part salt to 3 parts fish. Fermented in earthenware vats for 1–4 months. Final NaCl of 20–25% by weight. May be further pounded and coloured up with Angkak (a red colouring agent made from rice by action of mould *Monascus purpureus*). Pickle appearing at surface of fermenting mass removed and may be used as fish sauce. Proteolysis by autolytic enzymes releases peptides, amino acids, amines and ammonia. Minor role for salt-tolerant bacteria of Micrococcus, Staphylococcus, Pediococcus and Bacillus | East Asia, South East Asia |
| Basi | Alcoholic wine from sugar cane juice. Extracted by pressing cane, stored up to year, concentrated by boiling, leaves from guava may be immersed late in boiling. Filter into earthenware containers. Cooled to 40–45°C. Starter may be added, perhaps dried rotting fruit. 30–35°C, 4–6 days, or left 3–9 months. Starter comprises yeasts (Saccharomyces and Endomycopsis) and bacteria – lactic acid bacteria, especially Lactobacillus | East Asia, South East Asia, Africa |

**Table 18.1**   *Continued*

| Foodstuff | Details | Origin |
|---|---|---|
| Bongkrek | Coconut press cake, bound by mould mycelium into solid mass. Fried in oil and eaten with soup. Press cake remaining after coconut oil extract, for example, from copra is soaked for several hours in water. Vinegar may be added to lower pH. Pressed, sun-dried, steamed, cooled, inoculated with mould. Fermented on banana leaves, plastic sheets, mats or trays in dark, 24–48 h, 30–35°C. Mould mycelium penetrates and knits everything together. Mould *Rhizopus oligosporus* or *Neurospora sitophila* | South East Asia |
| Cachaça | Sugar cane spirit, 38+% alcohol | Brazil |
| Chicha | Effervescent sour alcoholic beverage. Yellow to red in colour made from maize or other starch crops, for example, cassava or beans. Dates to Inca. Chewed (normally women) but these days amylases may be developed via malting. Boiled with water, left 24 h to extract soluble materials, re-boiled. Sugars and molasses may be added. Filtered and the wort left to ferment in previously used containers. 20–30°C for 1–5 days. Lactic acid bacteria especially Lactobacillus, yeast, Acetobacter. Limit the life of the product to the time until which excess acetic acid is produced | South America |
| Corned beef (see also Chapter 13) | Usually from brisket – canned. Curing, but some mild fermentation. Name derives from large grains of salt used, which were called 'corns'. Beef salted in brine or pickle or the pickle is injected in more modern processes. Curing pickle sodium chloride, potassium or sodium nitrate or sodium nitrite, spices and herbs. These may include laurel, allspice, celery and onions. Placed in covered pickle for up to 2–3 weeks. Cooked in water or steamed to internal temperature of 68–71°C, cooled. May be canned and re-cooked. Micrococcus and some lactic acid bacteria | International |
| Country ham (see also Chapter 13) | Semi-dried cured pork. Salted and dried usually uncooked, may be smoked. Matured several months. For example, Cumberland, Kentucky, Parma (seasoned with pepper, allspice coriander and mustard and rubbed with pepper). Smithfield ham heavily smoked with hickory. Salts used are sodium chloride and potassium or sodium nitrate. Sometimes sugar used. Flavourings added to curing salt. Left at 5–15°C for 2–4 weeks and further pickling added, more weeks or months before cold smoking at 30–40°C over 1–5 weeks. Matured at 20–25°C for up to 2 years. Ham dries in this period. Nitrate to nitrite by Micrococcus and Staphylococcus. Some lactic acid bacteria, especially *Lactobacillus casei, Lactobacillus plantarum*. Some moulds especially *Penicillium nalgiovense* or Aspergillus spp. may coat surface of dried hams | International |
| Dried fish | Salted low-fat fish dried to various degrees. Storage and preservation in hot countries. Eviscerate and salt to 30–35% of weight with sodium chloride, loaded into barrels left at ambient (20–35°C) for 5–128 days. Removed from containers and sun- or air-dried for several weeks or even months. May be smoked in this period. Only salt-tolerant Micrococcus, Staphylococcus, Bacillus and lactic acid bacteria (Pediococcus and Lactobacillus) will survive | International |
| Dried meat (see also Chapter 13) | For example, salt beef, pastrami. Semi-dried uncooked meat (beef, lamb, goat, etc.) that has been cured, smoked and dried. Pieces of meat heavily salted with sodium chloride, potassium or sodium nitrate or sodium nitrite. Sugars, spices and seasonings. 5–15°C at high humidity (80–90% RH) at first, later high temperature and low humidity to encourage drying. | International |

**Table 18.1**  *Continued*

| Foodstuff | Details | Origin |
|---|---|---|
| | Cold smoking 32–38°C for 2–8 days before maturing for several weeks. Chemical curing with nitrates aided by Micrococcus and Staphylococcus reducing nitrate to nitrite. Also some fermentative lactic acid bacteria and yeast may develop. Pastrami (as an example) beef usually, black pepper, nutmeg, paprika, garlic and allspice. Smoked | |
| Fermented egg | Whole eggs (especially duck) coated in salt and ash paste and coated in rice hulls. The salt coating likely to comprise sodium chloride, sodium carbonates, tea leaves, calcium oxide and ash from burning grass. Eggs rolled over hull mixture, packed into earthenware or porcelain jars. Tightly sealed with mud and salt. 20–30°C for 15–50 days. Sodium hydroxide made from reaction of lime and sodium carbonate enters through eggshell and denatures and coagulates the egg protein, that is, a chemical as opposed to a microbial 'fermentation' | East Asia, South East Asia |
| Fish sauce | Brown salty liquid produced by breakdown of fish by fish enzymes. Small marine or fresh water fish, shrimps used whole, cereal (usually rice) added and koji. 1–2 parts salt to 5 parts fish. Packed into jars, concrete tanks or wooden vats. Left to ferment 20–35°C for 3–15 months. Liquid separated by filtration. Solid residues may be used to make Bagoong. Autolytic breakdown of fish protein. Sometimes fresh pineapple juice or koji added as source of proteinases. Trimethylamine and ammonia key products. Salt-tolerant Staphylococcus, Micrococcus and Bacillus may play a minor role in flavour development | East Asia, South East Asia, Europe |
| German salami (see also Chapter 13) | Dry, smoked uncooked sausage usually medium chopped and medium seasoning. Cold (−4 to −2°C) lean meat chopped and mixed with sodium chloride, potassium nitrate or sodium nitrite. Sodium ascorbate, spices, seasonings, sugar and sometimes glucono-δ-lactone. Pork fat chopped in. Stuffed at −4°C into casings or reformed collagen or artificial cellulose. Transferred to 'green room' where fermentation takes place at 20–32°C under high RH for 18–48 h if starter culture added. Or 5–9 days if not. Usually hot smoked to an internal temperature of 55–63°C, dried slowly at 15–24°C. Micrococcus and *Staphylococcus carnosus* important in early stages, converting nitrate to nitrite and stabilising colour. Pediococcus and Lactobacillus become dominant and may be added as starters | Germany |
| Ghee | Clarified butter, usually from cow, goat, buffalo or sheep. Keeps well without refrigeration. Butter, cream or kaffir heated to 110–140°C to melt and evaporate water. Filtered through muslin. Cooled to solidify. Antioxidants added. Lactic acid bacteria – Leuconostoc, Streptococci, Lactobacillus. Severe heating kills lactic acid flora | Indian subcontinent, Middle East, South East Asia, Africa |
| Jerky (see also Chapter 13) | Lean meat, salted and sun- or air dried in strips or thin sheets. Hot climates – dry product with good keeping properties. Snack or crumbled into soups or stews. Meat pieces salted with sodium chloride and perhaps nitrate. Left several days. Micrococci and Staphylococci reduce nitrate. Some development of lactic acid bacteria for flavour | America, Africa |

**Table 18.1**  *Continued*

| Foodstuff | Details | Origin |
|---|---|---|
| Kanji | Strong flavoured red alcoholic beverage made from beet juice or carrot. Refreshing. Usually consumed in hot weather. Roots peeled and shredded, 100 parts root, 5–6 parts salt, 3–4 parts mustard seed, 400–500 parts water. Ferment at 26–34°C for 4–7 days. Liquid drained for drinking. Portions of previous kanji may be added as a starter. *Hansenula anomala* and *Candida guilliermondii*, *Candida tropicalis* and *Geotrichum candidum* are active in fermentation | India, Israel |
| Kefir (see also Chapter 11) | Acidic and mildly alcohol effervescent milk from cows, buffalo goat milk. Heated to 90–95°C for 3–5 min. Cooled. Put into earthenware vessels. Inoculated with 5% kefir grains or 2–3% other starter. Ferment at 20–25°C for 10–24h, cooled to 12–16°C for a further 14–18 h, 'ripened' at 6–10°C for 5–8 days. Foamy and creamy. Diverse lactic acid bacteria: *Lactobacillus casei, Lactobacillus acidophilus, Steptococcus lactis*. Produce lactic acid from lactose. *Lactobacillus bulgaricus* produces acetaldehyde, *Leuconostoc cremoris* produce diacetyl and acetoin and *Lactobacillus brevis* makes acetoin, acetic acid, ethanol and $CO_2$. *Candida kefyr* and *Kluyveromyces fragilis* convert lactose to ethanol and $CO_2$ during the cooler ripening period | Middle East, Europe, North Africa |
| Kimchi (see also Chapter 15) | Mildly acidic carbonated vegetables – radish, Chinese cucumber, Chinese cabbage. Essential dish at most Korean meals. Vegetables mixed with small amounts of onion, chilli pepper, garlic, ginger or other flavouring agent and 4–6% salt or brine. Large earthenware vessels. Fish (shrimps, oysters) may be added to flavour. Left in a cool place to ferment often in cellar 10–18°C for 5–20 days. Maturation may be continued for many weeks if cool. Facultative lactic acid bacteria including *Leuconostoc mesenteroides, Streptococcus fecalis*, Pediococcus, *Lactobacillus plantarum, Lactobacillus brevis*. Aerobic bacteria Alcaligenes, Flavobacterium, Pseudomonas and *Bacillus megaterium* also grow. Later stages some yeast and moulds. Diverse organic acids | East Asia |
| Mead | Sweet alcoholic beverage from fermentation of honey with water or fruit juice. Often spiced. Honey added to 3–4 volumes of water or sometimes fruit juice often with addition of hops, herbs or spices. Usually boiled together. Surface froth skimmed off. 2–3% brewer's yeast added as starter. Ferment 15–25°C for 3–6 weeks. Usually aged in oak casks at 10–15°C for up to 10 years. Periodically transferred between casks or racked to remove deposits. Usually pasteurised, clarified and filtered. Lactic acid bacteria also involved – Lactobacilli with production of lactic and other compounds and lowering of pH | International |
| Nata | Thick white or cream-coloured gelatinous film growing on surface of juice from coconut, pineapple, sugar cane or other fruit waste. Eaten as dessert. Fruit juice mixture and pulp ground to a mash and diluted with water, 2% glacial acetic acid, 15% sucrose plus 0.5% ammonium dihydrogen phosphate. 10% inoculum of 48 h culture of acetic acid bacteria added to mixture in jars 28–31°C for 12–15 days. The thick layer of cells plus polysaccharide which forms on surface is washed to remove acetic acid, boiled and candied with 50% sucrose. Stored in barrels till needed. *Acetobacter aceti* ssp. Xylinum produces an extracellular polymer that can hold 25–30 times own water in gel | Philippines |

**Table 18.1**   *Continued*

| Foodstuff | Details | Origin |
|---|---|---|
| Papadum | Thin dried sheets of legume, cereal or starch crop flour. Stiff paste made by pounding legume flour, for example, *Phaseolus aureus* or Mung bean, *Phaseolus mungo*. Or rice flour, potato, sago or mix. Salt, spices including cardamom, caraway, pepper may be added. Dough made into long cylinder then portions cut and greased and rolled out very thinly. Ferment in sun for several hours. Usually stored in tins until needed. Served after baking in hot fire or deep-frying in oil. Saccharomyces, Candida and lactic acid bacteria all involved | Indian subcontinent |
| Pepperoni (see also Chapter 13) | Dried meat sausage – production closely similar to German salami. Moulds of *Penicillium nagliovense* and Aspergillus grow on surface and impact flavour | Europe, North America, Oceania |
| Pickled fish | Fatty fish, for example, herring pickled in salt sugar and acid brine. Up to 1.5 h. Usually whole or head removed, 15–17% salt, 5–7% sugar plus added spices and put in barrels. Left to ferment for several months 5–15°C. More salt and sugar may be added. After perhaps more than 1 year, fish washed and filleted and cut into pieces and packed in pickles of salt, sugar and acid (5–12% acetic). Proteolysis by cathepsins (endogenous proteinases). Softening of texture. Lactic acid bacteria of Pediococci, Leuconostoc, and Lactobacillus and salt-tolerant Micrococcus and Bacillus and yeasts play a minor role in flavour development | International |
| Pickled fruit | For example, cucumber, dill, but also lime pickle. Pick fruit under-ripe keeping sugar low and acidity high. Wash, dry, 2–3% salt or brine (5–10% salt). Sometimes inoculated with salt by needle. Herbs and spices may be added. Large earthenware jars filled, covered and sealed. 10–15°C for 2–6 weeks. Vinegar, salt and sugar may be added in modern commercial operations to replace traditional fermentation process. Gram-negative Enterobacter grow first, then lactic acid bacteria Leuconostoc, Streptococci, Pediococci, Lactobacillus dominate, producing lactic acid, acetic, ethanol, $CO_2$. Yeast then start to dominate, converting some of the acid to ethanol. If containers opened, oxidative growth occurs | International |
| Pisco | Distilled alcoholic beverage from South American wines | South America |
| Tea | Leaves and shoots of evergreen tree *Camellia sinensis*. Pruned to bush. Leaves rolled and fermented. Young leaves and shoots picked by hand. Wither 18–24 h, partly fermented. First for Oolong tea or for black tea, rolled directly, cells broken, release contents including enzymes and gives leaf a characteristic twist. Leaves spread in layers 10–15-mm deep in high humidity rooms to ferment 3–6 h. Colour goes from green to light brown. Fired by placing on trays through hot air (70–95°C) and colour goes dark brown. Sorted and classified and packed as dried tea. Black tea can be classified into top quality orange pekoe, from young shoots and leaf tips and souchong, medium quality and made from lower leaves. Green tea: fresh leaves are streamed to make them more pliable and to prevent fermentation, then rolled and fired. Oolong tea – leaves partially fermented before being dried. Fermentation primarily by enzymes released in rolling process. Especially oxidation. Perhaps minor role by bacteria and yeasts | East Asia, South East Asia, Indian subcontinent, Africa |

**Table 18.1**  *Continued*

| Foodstuff | Details | Origin |
|---|---|---|
| Tempe | Beans, mostly soy, bound together by mould mycelium into cake, sliced and dipped into soy or fish sauce or cooked in batter. Or in soups. Soybeans or other legume beans cleaned and soaked in water for 1–12 h. Some fermentation takes place. Then boiled for 1–3 h. Cooled, de-hulled, drained, inoculated with mould or a previous batch of tempe, wrapped in banana leaves or perforated polythene bags allowed to ferment at 27–32°C for 36–48 h. Mycelium penetrates. In initial soaking some early growth of Enterobacteriacea including *Klebsiella pneumoniae*, which makes Vitamin $B_{12}$, then lactic acid bacteria dominate, making lactic acid and lowering pH to 4.6–5.2. Helps establish mould *Rhizopus oligosporus* used in second stage. It releases proteinases. Ammonia produced, ergo pH rises again to 6.5–7. Some lipase released – with up to 25% of lipid converted to free fatty acids | South East Asia |
| Tequila | Mexican. Juice from *Agave tequilara* fermented by *Saccharomyces cerevisiae* and distilled and matured in oak | South America |
| Thickeners | Various microbially derived thickeners are now available to go alongside more traditional agents such as starch, pectins, alginates, plant gums and cellulose derivatives. Examples are xanthan (*Xanthomonas campestris* growing on glucose switches to gum production when the supply of nitrogen is depleted), gellan (*Pseudomonas elodea*), pullulan (*Aureobasidium pullulans*) | International |
| Vermouth | Fortified herb and spice-flavoured wine. Usually Muscat flavoured by mixing in approximately 0.5% of macerate of herbs and spices for 1–2 weeks. Daily mixing. When desired flavour reached, the wine is drawn off and filtered. Refrigerated and cold stored for >1 year. Now herb essences and extracts may be used. French vermouths lower in sugar content and higher in colour and alcohol when compared to Italian. Dry vermouths incorporate more wormwood and bitter orange peel, *Citrus auranticum* while sweet ones contain coriander, cinnamon, and cloves | Europe |
| Worcestershire sauce | Soybeans, anchovies, tamarinds, shallots, garlic, onion, salt, spices and flavouring added to vinegar, molasses and sugar. Allowed to ferment 4–6 months with occasional agitation. After maturation, the mix is pressed through a mesh screen that allows just the finer particles to pass. Pasteurised to stop fermentation, then bottled | England |

# Bibliography

Campbell-Platt, G. (1987) *Fermented Foods of the World: A Dictionary and Guide.* London: Butterworths.

# Index

CPSIA information can be obtained at www.ICGtesting.com
Printed in the USA
BVOW06*0415131213

338891BV00020B/508/P